Learning R

Richard Cotton

O'REILLY®

Beijing · Cambridge · Farnham · Köln · Sebastopol · Tokyo

Learning R

by Richard Cotton

Copyright © 2013 Richard Cotton. All rights reserved.

Published by O'Reilly Media, Inc., 1005 Gravenstein Highway North, Sebastopol, CA 95472.

O'Reilly books may be purchased for educational, business, or sales promotional use. Online editions are also available for most titles (*http://my.safaribooksonline.com*). For more information, contact our corporate/institutional sales department: 800-998-9938 or *corporate@oreilly.com*.

Editor: Meghan Blanchette
Production Editor: Kristen Brown
Copyeditor: Rachel Head
Proofreader: Jilly Gagnon

Indexer: WordCo Indexing Services
Cover Designer: Karen Montgomery
Interior Designer: David Futato
Illustrator: Rebecca Demarest

September 2013:   First Edition

Revision History for the First Edition:

2013-09-06:   First release

See *http://oreilly.com/catalog/errata.csp?isbn=9781449357108* for release details.

ISBN: 978-1-449-35710-8

[LSI]

Table of Contents

Part I.   The R Language

Part II.   The Data Analysis Workflow

Preface

R is a programming language and a software environment for data analysis and statistics. It is a GNU project, which means that it is free, open source software. It is growing exponentially by most measures—most estimates count over a million users, and it has over 4,000 add-on packages contributed by the community, with that number increasing by about 25% each year. The Tiobe Programming Community Index (*http://bit.ly/184JctZ*) of language popularity places it at number 24 at the time of this writing, roughly on a par with SAS and MATLAB.

R is used in almost every area where statistics or data analyses are needed. Finance, marketing, pharmaceuticals, genomics, epidemiology, social sciences, and teaching are all covered, as well as dozens of other smaller domains.

About This Book

Since R is primarily designed to let you do statistical analyses, many of the books written about R focus on teaching you how to calculate statistics or model datasets. This unfortunately misses a large part of the reality of analyzing data. Unless you are doing cutting-edge research, the statistical techniques that you use will often be routine, and the modeling part of your task may not be the largest one. The complete workflow for analyzing data looks more like this:

1. Retrieve some data.
2. Clean the data.
3. Explore and visualize the data.
4. Model the data and make predictions.
5. Present or publish your results.

Of course at each stage your results may generate interesting questions that lead you to look for more data, or for a different way to treat your existing data, which can send you back a step. The workflow can be iterative, but each of the steps needs to be undertaken.

The first part of this book is designed to teach you R from scratch—you don't need any experience in the language. In fact, no programming experience *at all* is necessary, but if you have some basic programming knowledge, it will help. For example, the book explains how to comment your code and how to write a `for` loop, but doesn't explain in great detail what they are. If you want a really introductory text on how to program, then *Python for Kids* by Jason R. Briggs is as good a place to start as any!

The second part of the book takes you through the complete data analysis workflow in R. Here, some basic statistical knowledge is assumed. For example, you should understand terms like *mean* and *standard deviation*, and what a bar chart is.

The book finishes with some more advanced R topics, like object-oriented programming and package creation. Garrett Grolemund's *Data Analysis with R* picks up where this book leaves off, covering data analysis workflow in more detail.

A word of warning: this isn't a reference book, and many of the topics aren't covered in great detail. This book provides tutorials to give you ideas about what you can do in R and let you practice. There isn't enough room to cover all 4,000 add-on packages, but by the time you've finished reading, you should be able to find the ones that you need, and get the help you need to start using them.

What Is in This Book

This is a book of two halves. The first half is designed to provide you with the technical skills you need to use R; each chapter is a short introduction to a different set of data types (for example, Chapter 4 covers vectors, matrices, and arrays) or a concept (for example, Chapter 8 covers branching and looping).

The second half of the book ramps up the fun: you get to see real data analysis in action. Each chapter covers a section of the standard data analysis workflow, from importing data to publishing your results.

Here's what you'll find in Part I, The R Language:

- Chapter 1, *Introduction*, tells you how to install R and where to get help.
- Chapter 2, *A Scientific Calculator*, shows you how to use R as a scientific calculator.
- Chapter 3, *Inspecting Variables and Your Workspace*, lets you inspect variables in different ways.
- Chapter 4, *Vectors, Matrices, and Arrays*, covers vectors, matrices, and arrays.

- Chapter 5, *Lists and Data Frames*, covers lists and data frames (for spreadsheet-like data).
- Chapter 6, *Environments and Functions*, covers environments and functions.
- Chapter 7, *Strings and Factors*, covers strings and factors (for categorical data).
- Chapter 8, *Flow Control and Loops*, covers branching (`if` and `else`), and basic looping.
- Chapter 9, *Advanced Looping*, covers advanced looping with the `apply` function and its variants.
- Chapter 10, *Packages*, explains how to install and use add-on packages.
- Chapter 11, *Dates and Times*, covers dates and times.

Here are the topics covered in Part II, The Data Analysis Workflow:

- Chapter 12, *Getting Data*, shows you how to import data into R.
- Chapter 13, *Cleaning and Transforming*, explains cleaning and manipulating data.
- Chapter 14, *Exploring and Visualizing*, lets you explore data by calculating statistics and plotting.
- Chapter 15, *Distributions and Modeling*, introduces modeling.
- Chapter 16, *Programming*, covers a variety of advanced programming techniques.
- Chapter 17, *Making Packages*, shows you how to package your work for others.

Lastly, there are useful references in Part III, Appendixes:

- Appendix A, *Properties of Variables*, contains tables comparing the properties of different types of variables.
- Appendix B, *Other Things to Do in R*, describes some other things that you can do in R.
- Appendix C, *Answers to Quizzes*, contains the answers to the end-of-chapter quizzes.
- Appendix D, *Solutions to Exercises*, contains the answers to the end of chapter programming exercises.

Which Chapters Should I Read?

If you have never used R before, then start at the beginning and work through chapter by chapter. If you already have some experience with R, you may wish to skip the first chapter and skim the chapters on the R core language.

Each chapter deals with a different topic, so although there is a small amount of dependency from one chapter to the next, it is possible to pick and choose chapters that interest you.

I recently discussed this matter with Andrie de Vries, author of *R For Dummies*. He suggested giving up and reading his book instead![1]

Conventions Used in This Book

The following font conventions are used in this book:

Italic
> Indicates new terms, URLs, email addresses, file and pathnames, and file extensions.

`Constant width`
> Used for code samples that should be copied verbatim, as well as within paragraphs to refer to program elements such as variable or function names, data types, environment variables, statements, and keywords. Output from blocks of code is also in constant width, preceded by a double hash (##).

`Constant width italic`
> Shows text that should be replaced with user-supplied values or by values determined by context.

There is a style guide for the code used in this book at *http://4dpiecharts.com/r-code-style-guide*.

> This icon signifies a tip, suggestion, or general note.

> This icon indicates a warning or caution.

Goals, Summaries, Quizzes, and Exercises

Each chapter begins with a list of goals to let you know what to expect in the forthcoming pages, and finishes with a summary that reiterates what you've learned. You also get a quiz, to make sure you've been concentrating (and not just pretending to read while watching telly). The answers to the questions can be found within the chapter (or at the

---

1. Andrie's book covers much the same ground as *Learning R*, and in many ways is almost as good as this work, so I won't be offended if you want to read it too.

end of the book, if you want to cheat). Finally, each chapter concludes with some exercises, most of which involve you writing some R code. After each exercise description there is a number in square brackets, denoting a generous estimate of how many minutes it might take you to complete it.

Using Code Examples

Supplemental material (code examples, exercises, etc.) is available for download at *http://cran.r-project.org/web/packages/learningr*.

This book is here to help you get your job done. In general, if example code is offered with this book, you may use it in your programs and documentation. You do not need to contact us for permission unless you're reproducing a significant portion of the code. For example, writing a program that uses several chunks of code from this book does not require permission. Selling or distributing a CD-ROM of examples from O'Reilly books does require permission. Answering a question by citing this book and quoting example code does not require permission. Incorporating a significant amount of example code from this book into your product's documentation does require permission.

We appreciate, but do not require, attribution. An attribution usually includes the title, author, publisher, and ISBN. For example: "*Learning R* by Richard Cotton (O'Reilly). Copyright 2013 Richard Cotton, 978-1-449-35710-8."

If you feel your use of code examples falls outside fair use or the permission given above, feel free to contact us at *permissions@oreilly.com*.

Safari® Books Online

 *Safari Books Online* is an on-demand digital library that delivers expert content in both book and video form from the world's leading authors in technology and business.

Technology professionals, software developers, web designers, and business and creative professionals use Safari Books Online as their primary resource for research, problem solving, learning, and certification training.

Safari Books Online offers a range of product mixes and pricing programs for organizations, government agencies, and individuals. Subscribers have access to thousands of books, training videos, and prepublication manuscripts in one fully searchable database from publishers like O'Reilly Media, Prentice Hall Professional, Addison-Wesley Professional, Microsoft Press, Sams, Que, Peachpit Press, Focal Press, Cisco Press, John Wiley & Sons, Syngress, Morgan Kaufmann, IBM Redbooks, Packt, Adobe Press, FT Press, Apress, Manning, New Riders, McGraw-Hill, Jones & Bartlett, Course Technology, and dozens more. For more information about Safari Books Online, please visit us online.

How to Contact Us

Please address comments and questions concerning this book to the publisher:

O'Reilly Media, Inc.
1005 Gravenstein Highway North
Sebastopol, CA 95472
800-998-9938 (in the United States or Canada)
707-829-0515 (international or local)
707-829-0104 (fax)

We have a web page for this book, where we list errata, examples, and any additional information. You can access this page at *http://oreil.ly/learningR*.

To comment or ask technical questions about this book, send email to *bookques tions@oreilly.com*.

For more information about our books, courses, conferences, and news, see our website at *http://www.oreilly.com*.

Find us on Facebook: *http://facebook.com/oreilly*

Follow us on Twitter: *http://twitter.com/oreillymedia*

Watch us on YouTube: *http://www.youtube.com/oreillymedia*

Acknowledgments

Many amazing people have helped with the making of this book, not least my excellent editor Meghan Blanchette, who is full of sensible advice.

Data was donated by several wonderful people:

- Bill Hogan of AMD found and cleaned the Alpe d'Huez cycling dataset, and pointed me toward the CDC gonorrhoea dataset. He wanted me to emphasize that he's disease-free, ladies.
- Ewan Hunter of CEFAS provided the North Sea crab dataset.
- Corina Logan of the University of Cambridge compiled and provided the deer skull data.
- Edwin Thoen of Leiden University compiled and provided the Obama vs. McCain dataset.
- Gwern Branwen compiled the hafu dataset by watching and reading an inordinate amount of manga. Kudos.

Many other people sent me datasets; there wasn't room for them all, but thank you anyway!

Bill Hogan also reviewed the book, as did Daisy Vincent of Marin Software, and JD Long. I don't know where JD works, but he lives in Bermuda, so it probably involves triangles. Additional comments and feedback were provided by James White, Ben Hanks, Beccy Smith, and Guy Bourne of TDX Group; Alex Hogg and Adrian Kelsey of HSL; Tom Hull, Karen Vanstaen, Rachel Beckett, Georgina Rimmer, Ruth Wortham, Bernardo Garcia-Carreras, and Joana Silva of CEFAS; Tal Galili of Tel Aviv University; Garrett Grolemund of RStudio; and John Verzani of the City University of New York. David Maxwell of CEFAS wonderfully recruited more or less everyone else in CEFAS to review my book.

John Verzani also deserves much credit for helping conceive this book, and for providing advice on the structure.

Sanders Kleinfeld of O'Reilly provided great tech support when I was pulling my hair out over character encodings in the manuscript. Yihui Xie went above and beyond the call of duty helping me get knitr to generate AsciiDoc. Rachel Head single-handedly spotted over 4,000 bugs, typos, and mistakes while copyediting.

Garib Murshudov was the lecturer who first taught me R, back in 2004.

Finally, Janette Bowler deserves a medal for her endless patience and support while I've been busy writing.

The R Language

## CHAPTER 1

Introduction

Congratulations! You've just begun your quest to become an R programmer. So you don't pull any mental muscles, this chapter starts you off gently with a nice warm-up. Before you begin coding, we're going to talk about what R is, and how to install it and begin working with it. Then you'll try writing your first program and learn how to get help.

Chapter Goals

After reading this chapter, you should:

- Know some things that you can use R to do
- Know how to install R and an IDE to work with it
- Be able to write a simple program in R
- Know how to get help in R

What Is R?

Just to confuse you, R refers to two things. There is R, the programming language, and R, the piece of software that you use to run programs written in R. Fortunately, most of the time it should be clear from the context which R is being referred to.

R (the language) was created in the early 1990s by Ross Ihaka and Robert Gentleman, then both working at the University of Auckland. It is based upon the S language that was developed at Bell Laboratories in the 1970s, primarily by John Chambers. R (the software) is a GNU project, reflecting its status as important free and open source software. Both the language and the software are now developed by a group of (currently) 20 people known as the R Core Team.

The fact that R's history dates back to the 1970s is important, because it has evolved over the decades, rather than having been designed from scratch (contrast this with, for example, Microsoft's .NET Framework, which has a much more "created"[1] feel). As with life-forms, the process of evolution has led to some quirks and inconsistencies. The upside of the more free-form nature of R (and the free license in particular) is that if you don't like how something in R is done, you can write a package to make it do things the way that you want. Many people have already done that, and the common question now is not "Can I do this in R?" but "Which of the three implementations should I use?"

R is an interpreted language (sometimes called a scripting language), which means that your code doesn't need to be compiled before you run it. It is a high-level language in that you don't have access to the inner workings of the computer you are running your code on; everything is pitched toward helping you analyze data.

R supports a mixture of programming paradigms. At its core, it is an imperative language (you write a script that does one calculation after another), but it also supports object-oriented programming (data and functions are combined inside classes) and functional programming (functions are *first-class objects*; you treat them like any other variable, and you can call them recursively). This mix of programming styles means that R code can bear a lot of similarity to several other languages. The curly braces mean that you can write imperative code that looks like C (but the vectorized nature of R that we'll discuss in Chapter 2 means that you have fewer loops). If you use reference classes, then you can write object-oriented code that looks a bit like C# or Java. The functional programming constructs are Lisp-inspired (the variable-scoping rules are taken from the Lisp dialect, Scheme), but there are fewer brackets. All this is a roundabout way of saying that R follows the Perl ethos (*http://bit.ly/148zbcF*):

> There is more than one way to do it.
>
> — Larry Wall

Installing R

If you are using a Linux machine, then it is likely that your package manager will have R available, though possibly not the latest version. For everyone else, to install R you must first go to *http://www.r-project.org*. Don't be deceived by the slightly archaic website;[2] it doesn't reflect on the quality of R. Click the link that says "download R" (*http://cran.r-project.org/mirrors.html*) in the "Getting Started" pane at the bottom of the page.

---

1. Intelligently designed?

2. A look in the Internet Archive's Wayback Machine suggests that the front page hasn't changed much since May 2004. (*http://web.archive.org/web/20040415000000*/*http://www.r-project.org/*)

Once you've chosen a mirror close to you, choose a link in the "Download and Install R" pane at the top of the page that's appropriate to your operating system. After that there are one or two OS-specific clicks that you need to make to get to the download.

If you are a Windows user who doesn't like clicking, there is a cheeky shortcut to the setup file at *http://<CRAN MIRROR>/bin/windows/base/release.htm*.

Choosing an IDE

If you use R under Windows or Mac OS X, then a graphical user interface (GUI) is available to you. This consists of a command-line interpreter, facilities for displaying plots and help pages, and a basic text editor. It is perfectly possible to use R in this way, but for serious coding you'll at least want to use a more powerful text editor. There are countless text editors for programmers; if you already have a favorite, then take a look to see if you can get syntax highlighting of R code for it.

If you aren't already wedded to a particular editor, then I suggest that you'll get the best experience of R by using an integrated development environment (IDE). Using an IDE rather than a separate text editor gives you the benefit of only using one piece of software rather than two. You get all the facilities of the stock R GUI, but with a better editor, and in some cases things like integrated version control.

The following sections introduce five popular choices, but this is by no means an exhaustive list (a few additional suggestions follow). It is worth trying several IDEs; a development environment is a piece of software that you could be spending thousands of hours using, so it's worth taking the time to find one[3] that you like. A few additional suggestions follow this selection.

Emacs + ESS

Although Emacs calls itself a text editor, 36 years (and counting) of development have given it an unparalleled number of features. If you've been programming for any substantial length of time, you probably already know whether or not you want to use it. Converts swear by its limitless customizability and raw editing power; others complain that it overcomplicates things and that the key chords give them repetitive strain injury. There is certainly a steep learning curve, so be willing to spend a month or two getting used to it. The other big benefit is that Emacs is not R-specific, so you can use it for programming in many languages. The original version of Emacs is (like R) a GNU project, available from *http://www.gnu.org/software/emacs/*.

---

3. You don't need to limit yourself to just *one* way of using R. I have IDE commitment issues and use a mix of Eclipse + StatET, RStudio, Live-R, Tinn-R, Notepad++, and R GUI. Experiment, and find something that works for you.

Another popular fork is XEmacs, available from *http://www.xemacs.org/*.

Emacs Speaks Statistics (ESS) is an add-on for Emacs that assists you in writing R code. Actually, it works with S-Plus, SAS, and Stata, too, so you can write statistical code with whichever package you like (choose R!). Several of the authors of ESS are also R Core Team members, so you are guaranteed good integration with R. It is available through the Emacs package management system, or you can download it from *http://ess.r-project.org/*.

Use it if you want to write code in multiple languages, you want the most powerful editor available, and you are fearless with learning curves.

Eclipse/Architect

Eclipse is another cross-platform IDE, widely used in the Java community. Like Emacs, it is very powerful, and its plug-in system makes it highly customizable. The learning curve is shallower, though, and it allows for more pointing and clicking than the heavily keyboard-driven Emacs.

Architect is an R-oriented variant of Eclipse developed by statistics consultancy Open Analytics. It includes the StatET plug-in for integration with R, including a debugger that is superior to the one built into R GUI. Download it from *http://www.openanalyt ics.eu/downloads/architect*.

Alternatively, you can get the standard Eclipse IDE from *http://eclipse.org* and use its package manager to download the StatET plug-in from *http://www.walware.de/goto/ statet*.

Use it if you want to write code in multiple languages, you don't have time to learn Emacs, and you don't mind a several-hundred-megabyte install.

RStudio

RStudio is an R-specific IDE. That means that you lose the ability to code (easily) in multiple languages, but you do get some features especially for R. For example, the plot windows are better than the R GUI originals, and there are facilities for publishing code. The editor is more basic than either Emacs or Eclipse, but it's good enough for most purposes, and is easier to get started with than the other two. RStudio's party trick is that you can run it remotely through a browser, so you can run R on a powerful server, then access it from a netbook (or smartphone) without loss of computational power. Download it from *http://www.rstudio.org*.

Use it if you mainly write R code, don't need advanced editor features, and want a shallow learning curve or the ability to run remotely.

Revolution-R

Revolution-R comes in two flavors: the free (as in beer) community edition and the paid-for enterprise edition. Both take a different tack from the other IDEs mentioned so far: whereas Emacs, Eclipse, and RStudio are pure graphical frontends that let you connect to any version of R, Revolution-R ships with its own customized version of R. Typically this is a stable release, one or two versions back from the most current. It also has some enhancements for working with big data, and some enterprise-related features. Download it from *http://www.revolutionanalytics.com/products/revolution-r.php*.

Use it if you mainly write R code, you work with big data or want a paid support contract, or you require extra stability in your R platform.

Live-R

Live-R is a new player, in invite-only beta at the time this book is going to press. It provides an IDE for R as a web application. This avoids all the hassle of installing software on your machine and, like RStudio's remote installation, gives you the ability to run R calculations from an underpowered machine. Live-R also includes a number of features for collaboration, including a shared editor and code publishing, as well as some admin tools for running courses based upon R. The main downside is that not all the add-on packages for R are available; you are currently limited to about 200 or so that are compatible with the web application. Sign up at *http://live-analytics.com/*.

Use it if you mainly write R code, don't want to install any software, or want to teach a class based upon R.

Other IDEs and Editors

There are many more editors that you can use to write R code. Here's a quick roundup of a few more possibilities:

- JGR (*http://rforge.net/JGR*) [pronounced "Jaguar"] is a Java-based GUI for R, essentially a souped-up version of the stock R GUI.
- Tinn-R (*http://www.sciviews.org/Tinn-R*) is a fork of the editor TINN that has extensions specifically to help you write R code.
- SciViews-K (*http://www.sciviews.org/SciViews-K*), from the same team that makes Tinn-R, is an extension for the Komodo IDE to work with R.
- Vim-R (*http://www.vim.org/scripts/script.php?script_id=2628*) is a plug-in for Vim that provides R integration.
- NppToR (*http://sourceforge.net/projects/npptor*) plugs into Notepad++ to give R integration.

Your First Program

It is a law of programming books that the first example shall be a program to print the phrase "Hello world!" In R that's really boring, since you just type "Hello world!" at the command prompt, and it will parrot it back to you. Instead, we're going to write the simplest statistical program possible.

Open up R GUI, or whichever IDE you've decided to use, find the command prompt (in the code editor window), and type:

```
mean(1:5)
```

Hit Enter to run the line of code. Hopefully, you'll get the answer 3. As you might have guessed, this code is calculating the arithmetic mean of the numbers from 1 to 5. The colon operator, :, creates a sequence of numbers from the first number, in this case 1, to the second number (5), each separated by 1. The resulting sequence is called a *vector*. mean is a *function* (that calculates the arithmetic mean), and the vector that we enclose inside the parentheses is called an *argument* to the function.

Well done! You've calculated a statistic using R.

 In R GUI and most of the IDEs mentioned here, you can press the up arrow key to cycle back through previous commands.

How to Get Help in R

Before you get started writing R code, the most important thing to know is how to get help. There are lots of ways to do this. Firstly, if you want help on a function or a dataset that you know the name of, type ? followed by the name of the function. To find functions, type two question marks (??) followed by a keyword related to the problem to search. Special characters, reserved words, and multiword search terms need enclosing in double or single quotes. For example:

```
?mean                 #opens the help page for the mean function
?"+"                  #opens the help page for addition
?"if"                 #opens the help page for if, used for branching code
??plotting            #searches for topics containing words like "plotting"
??"regression model"  #searches for topics containing phrases like this
```

 That # symbol denotes a comment. It means that R will ignore the rest of the line. Use comments to document your code, so that you can remember what you were doing six months ago.

The functions `help` and `help.search` do the same things as ? and ??, respectively, but with these you always need to enclose your arguments in quotes. The following commands are equivalent to the previous lot:

```
help("mean")
help("+")
help("if")
help.search("plotting")
help.search("regression model")
```

The `apropos` function[4] finds variables (including functions) that match its input. This is really useful if you can only half-remember the name of a variable that you've created, or a function that you want to use. For example, suppose you create a variable a_vector:

```
a_vector <- c(1, 3, 6, 10)
```

You can then recall this variable using `apropos`:

```
apropos("vector")
## [1] "._C__vector"    "a_vector"         "as.data.frame.vector"
## [4] "as.vector"      "as.vector.factor" "is.vector"
## [7] "vector"         "Vectorize"
```

The results contain the variable you just created, a_vector, and all other variables that contain the string `vector`. In this case, all the others are functions that are built into R.

Just finding variables that contain a particular string is fine, but you can also do fancier matching with `apropos` using regular expressions.

Regular expressions are a cross-language syntax for matching strings. The details will only be touched upon in this book, but you need to learn to use them; they'll change your life. Start at *http://www.regular-expressions.info/quickstart.html*, and then try Michael Fitzgerald's *Introducing Regular Expressions*.

A simple usage of `apropos` could, for example, find all variables that end in z, or to find all variables containing a number between 4 and 9:

```
apropos("z$")
## [1] "alpe_d_huez" "alpe_d_huez" "force_tz"  "indexTZ"   "SSgompertz"
## [6] "toeplitz"    "tz"          "unz"       "with_tz"
```

---

4. `apropos` is Latin for "A Unix program that finds manpages."

```
apropos("[4-9]")

##  [1] "._C__S4"           "._T__xmlToS4:XML"    ".parseISO8601"
##  [4] ".SQL92Keywords"    ".TAOCP1997init"      "asS4"
##  [7] "assert_is_64_bit_os" "assert_is_S4"       "base64"
## [10] "base64Decode"      "base64Encode"        "blues9"
## [13] "car90"             "enc2utf8"            "fixPre1.8"
## [16] "Harman74.cor"      "intToUtf8"           "is_64_bit_os"
## [19] "is_S4"             "isS4"                "seemsS4Object"
## [22] "state.x77"         "to.minutes15"        "to.minutes5"
## [25] "utf8ToInt"         "xmlToS4"
```

Most functions have examples that you can run to get a better idea of how they work. Use the example function to run these. There are also some longer demonstrations of concepts that are accessible with the demo function:

```
example(plot)
demo()           #list all demonstrations
demo(Japanese)
```

R is modular and is split into *packages* (more on this later), some of which contain *vignettes*, which are short documents on how to use the packages. You can browse all the vignettes on your machine using browseVignettes:

```
browseVignettes()
```

You can also access a specific vignette using the vignette function (but if your memory is as bad as mine, using browseVignettes combined with a page search is easier than trying to remember the name of a vignette and which package it's in):

```
vignette("Sweave", package = "utils")
```

The help search operator ?? and browseVignettes will only find things in packages that you have installed on your machine. If you want to look in *any* package, you can use RSiteSearch, which runs a query at *http://search.r-project.org*. Multiword terms need to be wrapped in braces:

```
RSiteSearch("{Bayesian regression}")
```

 Learning to help yourself is extremely important. Think of a keyword related to your work and try ?, ??, apropos, and RSiteSearch with it.

There are also lots of R-related resources on the Internet that are worth trying. There are too many to list here, but start with these:

- R has a number of mailing lists (*http://www.r-project.org/mail.html*) with archives containing years' worth of questions on the language. At the very least, it is worth signing up to the general-purpose list, *R-help*.

- RSeek (*http://rseek.org*) is a web search engine for R that returns functions, posts from the R mailing list archives, and blog posts.
- R-bloggers (*http://www.r-bloggers.com*) is the main R blogging community, and the best way to stay up to date with news and tips about R.
- The programming question and answer site Stack Overflow (*http://www.stackover flow.com*) also has a vibrant R community, providing an alternative to the *R-help* mailing list. You also get points and badges for answering questions!

Installing Extra Related Software

There are a few other bits of software that R can use to extend its functionality. Under Linux, your package manager should be able to retrieve them. Under Windows, rather than hunting all over the Internet to track down this software, you can use the installr add-on package to automatically install these extra pieces of software. None of this software is compulsory, so you can skip this section now if you want, but it's worth knowing that the package exists when you come to need the additional software. Installing and loading packages is discussed in detail in Chapter 10, so don't worry if you don't understand the commands yet:

```
install.packages("installr")   #download and install the package named installr
library(installr)               #load the installr package
install.RStudio()               #download and install the RStudio IDE
install.Rtools()                #Rtools is needed for building your own packages
install.git()                   #git provides version control for your code
```

Summary

- R is a free, open source language for data analysis.
- It's also a piece of software used to run programs written in R.
- You can download R from *http://www.r-project.org*.
- You can write R code in any text editor, but there are several IDEs that make development easier.
- You can get help on a function by typing ? then its name.
- You can find useful functions by typing ?? then a search string, or by calling the apropos function.
- There are many online resources for R.

Test Your Knowledge: Quiz

Question 1-1

Which language is R an open source version of?

Question 1-2

Name at least two programming paradigms in which you can write R code.

Question 1-3

What is the command to create a vector of the numbers from 8 to 27?

Question 1-4

What is the name of the function used to search for help within R?

Question 1-5

What is the name of the function used to search for R-related help on the Internet?

Test Your Knowledge: Exercises

Exercise 1-1

Visit *http://www.r-project.org*, download R, and install it. For extra credit, download and install one of the IDEs mentioned in "Other IDEs and Editors" on page 7. [30]

Exercise 1-2

The function `sd` calculates the standard deviation. Calculate the standard deviation of the numbers from 0 to 100. Hint: the answer should be about `29.3`. [5]

Exercise 1-3

Watch the demonstration on mathematical symbols in plots, using `demo(plot math)`. [5]

## CHAPTER 2

A Scientific Calculator

R is at heart a supercharged scientific calculator, so it has a fairly comprehensive set of mathematical capabilities built in. This chapter will take you through the arithmetic operators, common mathematical functions, and relational operators, and show you how to assign a value to a variable.

Chapter Goals

After reading this chapter, you should:

- Be able to use R as a scientific calculator
- Be able to assign a variable and view its value
- Be able to use infinite and missing values
- Understand what logical vectors are and how to manipulate them

Mathematical Operations and Vectors

The + operator performs addition, but it has a special trick: as well as adding two numbers together, you can use it to add two vectors. A *vector* is an ordered set of values. Vectors are tremendously important in statistics, since you will usually want to analyze a whole dataset rather than just one piece of data.

The colon operator, :, which you have seen already, creates a sequence from one number to the next, and the c function concatenates values, in this case to create vectors (*concatenate* is a Latin word meaning "connect together in a chain").

Variable names are case sensitive in R, so we need to be a bit careful in this next example. The C function does something completely different to c:[1]

```
1:5 + 6:10          #look, no loops!
## [1]  7  9 11 13 15
c(1, 3, 6, 10, 15) + c(0, 1, 3, 6, 10)
## [1]  1  4  9 16 25
```

 The colon operator and the c function are used almost everywhere in R code, so it's good to practice using them. Try creating some vectors of your own now.

If we were writing in a language like C or Fortran, we would need to write a loop to perform addition on all the elements in these vectors. The vectorized nature of R's addition makes things easy, letting us avoid the loop. Vectors will be discussed more in "Logical Vectors" on page 20.

Vectorized has several meanings in R, the most common of which is that an operator or a function will act on each element of a vector without the need for you to explicitly write a loop. (This built-in implicit looping over elements is also much faster than explicitly writing your own loop.) A second meaning of vectorization is when a function takes a vector as an input and calculates a summary statistic:

```
sum(1:5)
## [1] 15
median(1:5)
## [1] 3
```

A third, much less common case of vectorization is *vectorization over arguments*. This is when a function calculates a summary statistic from several of its input arguments. The sum function does this, but it is very unusual. median does not:

```
sum(1, 2, 3, 4, 5)
## [1] 15
median(1, 2, 3, 4, 5)  #this throws an error
## Error: unused arguments (3, 4, 5)
```

---

1. There are a few other name clashes: filter and Filter, find and Find, gamma and Gamma, nrow/ncol and NROW/NCOL. This is an unfortunate side effect of R being an evolved rather than a designed language.

All the arithmetic operators in R, not just plus (+), are vectorized. The following examples demonstrate subtraction, multiplication, exponentiation, and two kinds of division, as well as remainder after division:

```
c(2, 3, 5, 7, 11, 13) - 2          #subtraction
## [1]  0  1  3  5  9 11

-2:2 * -2:2                         #multiplication
## [1] 4 1 0 1 4

identical(2 ^ 3, 2 ** 3)           #we can use ^ or ** for exponentiation
                                   #though ^ is more common
## [1] TRUE

1:10 / 3                           #floating point division
##  [1] 0.3333 0.6667 1.0000 1.3333 1.6667 2.0000 2.3333 2.6667 3.0000 3.3333

1:10 %/% 3                         #integer division
##  [1] 0 0 1 1 1 2 2 2 3 3

1:10 %% 3                          #remainder after division
##  [1] 1 2 0 1 2 0 1 2 0 1
```

R also contains a wide selection of mathematical functions. You get trigonometry (sin, cos, tan, and their inverses asin, acos, and atan), logarithms and exponents (log and exp, and their variants log1p and expm1 that calculate $\log(1 + x)$ and $\exp(x - 1)$ more accurately for very small values of x), and almost any other mathematical function you can think of. The following examples provide a hint of what is on offer. Again, notice that all the functions naturally operate on vectors rather than just single values:

```
cos(c(0, pi / 4, pi / 2, pi))      #pi is a built-in constant
## [1]  1.000e+00  7.071e-01  6.123e-17 -1.000e+00

exp(pi * 1i) + 1                   #Euler's formula
## [1] 0+1.225e-16i

factorial(7) + factorial(1) - 71 ^ 2 #5041 is a great number
## [1] 0

choose(5, 0:5)
## [1]  1  5 10 10  5  1
```

To compare integer values for equality, use ==. Don't use a single = since that is used for something else, as we'll see in a moment. Just like the arithmetic operators, == and the other relational operators are vectorized. To check for inequality, the "not equals" operator is !=. Greater than and less than are as you might expect: > and < (or >= and <= if equality is allowed). Here are a few examples:

```
c(3, 4 - 1, 1 + 1 + 1) == 3                #operators are vectorized too
## [1] TRUE TRUE TRUE
1:3 != 3:1
## [1]  TRUE FALSE  TRUE
exp(1:5) < 100
## [1]  TRUE  TRUE  TRUE  TRUE FALSE
(1:5) ^ 2 >= 16
## [1] FALSE FALSE FALSE  TRUE  TRUE
```

Comparing nonintegers using == is problematic. All the numbers we have dealt with so far are floating point numbers. That means that they are stored in the form $a * 2 ^ b$, for two numbers a and b. Since this whole form has to be stored in 32 bits, the resulting number is only an approximation of what you really want. This means that rounding errors often creep into calculations, and the answers you expected can be wildly wrong. Whole books have been written on this subject; there is too much to worry about here. Since this is such a common mistake, the FAQ on R has an entry about it (*http://bit.ly/ 17jZFfE*), and it's a good place to start if you want to know more.

Consider these two numbers, which should be the same:

```
sqrt(2) ^ 2 == 2           #sqrt is the square-root function
## [1] FALSE
sqrt(2) ^ 2 - 2            #this small value is the rounding error
## [1] 4.441e-16
```

R also provides the function all.equal for checking equality of numbers. This provides a tolerance level (by default, about 1.5e-8), so that rounding errors less than the tolerance are ignored:

```
all.equal(sqrt(2) ^ 2, 2)
## [1] TRUE
```

If the values to be compared are not the same, all.equal returns a report on the differences. If you require a TRUE or FALSE value, then you need to wrap the call to all.equal in a call to isTRUE:

```
all.equal(sqrt(2) ^ 2, 3)
## [1] "Mean relative difference: 0.5"
isTRUE(all.equal(sqrt(2) ^ 2, 3))
## [1] FALSE
```

To check that two numbers are the same, don't use ==. Instead, use the `all.equal` function.

We can also use == to compare strings. In this case the comparison is case sensitive, so the strings must match exactly. It is also theoretically possible to compare strings using greater than or less than (> and <):

```
c(
  "Can", "you", "can", "a", "can", "as",
  "a", "canner", "can", "can", "a", "can?"
) == "can"
```

```
## [1] FALSE FALSE  TRUE FALSE  TRUE FALSE FALSE FALSE  TRUE  TRUE FALSE
## [12] FALSE
```

```
c("A", "B", "C", "D") < "C"
```

```
## [1]  TRUE  TRUE FALSE FALSE
```

```
c("a", "b", "c", "d") < "C" #your results may vary
```

```
## [1]  TRUE  TRUE  TRUE FALSE
```

In practice, however, the latter approach is almost always an awful idea, since the results depend upon your locale (different cultures are full of odd sorting rules for letters; in Estonian, "z" comes between "s" and "t"). More powerful string matching functions will be discussed in "Cleaning Strings" on page 191.

The help pages ?Arithmetic, ?Trig, ?Special, and ?Comparison have more examples, and explain the gory details of what happens in edge cases. (Try 0 ^ 0 or integer division on nonintegers if you are curious.)

Assigning Variables

It's all very well calculating things, but most of the time we want to store the results for reuse. We can assign a (local) variable using either <- or =, though for historical reasons, <- is preferred:

```
x <- 1:5
y = 6:10
```

Now we can reuse these values in our further calculations:

```
x + 2 * y - 3
```

```
## [1] 10 13 16 19 22
```

Notice that we didn't have to declare what types of variables x and y were going to be before we assigned them (unlike in most compiled languages). In fact, we *couldn't* have declared the type, since no such concept exists in R.

Variable names can contain letters, numbers, dots, and underscores, but they can't start with a number, or a dot followed by a number (since that looks too much like a number). Reserved words like "if" and "for" are not allowed. In some locales, non-ASCII letters are allowed, but for code portability it is better to stick to "a" to "z" (and "A" to "Z"). The help page ?make.names gives precise details about what is and isn't allowed.

The spaces around the assignment operators aren't compulsory, but they help readability, especially with <-, so we can easily distinguish assignment from less than:

```
x <- 3
x < -3
x<-3      #is this assignment or less than?
```

We can also do global assignment using <<-. There'll be more on what this means when we cover environments and scoping in "Environments" on page 79 in Chapter 6; for now, just think of it as creating a variable available anywhere:

```
x <<- exp(exp(1))
```

There is one more method of variable assignment, via the assign function. It is much less common than the other methods, but very occasionally it is useful to have a function syntax for assigning variables. Local ("normal") assignment takes two arguments—the name of the variable to assign to and the value you want to give it:

```
assign("my_local_variable", 9 ^ 3 + 10 ^ 3)
```

Global assignment (like the <<- operator does) takes an extra argument:

```
assign("my_global_variable", 1 ^ 3 + 12 ^ 3, globalenv())
```

Don't worry about the globalenv bit for now; as with scoping, it will be explained in Chapter 6.

> Using the assign function makes your code less readable compared to <-, so you should use it sparingly. It occasionally makes things easier in some advanced programming cases involving environments, but if your code is filled with calls to assign, you are probably doing something wrong.
>
> Also note that the assign function doesn't check its first argument to see if it is a valid variable name: it always just creates it.

Notice that when you assign a variable, you don't see the value that has been given to it. To see what value a variable contains, simply type its name at the command prompt to print it:

```
x
## [1] 1 2 3 4 5
```

 Under some systems, for example running R from a Linux terminal, you may have to explicitly call the print function to see the value. In this case, type print(x).

If you want to assign a value and print it all in one line, you have two possibilities. Firstly, you can put multiple statements on one line by separating them with a semicolon, ;. Secondly, you can wrap the assignment in parentheses, (). In the following examples, rnorm generates random numbers from a normal distribution, and rlnorm generates them from a lognormal distribution:[2]

```
z <- rnorm(5); z
## [1]  1.8503 -0.5787 -1.4797 -0.1333 -0.2321
(zz <- rlnorm(5))
## [1] 1.0148 4.2476 0.3574 0.2421 0.3163
```

Special Numbers

To help with arithmetic, R supports four special numeric values: Inf, -Inf, NaN, and NA. The first two are, of course, positive and negative infinity, but the second pair need a little more explanation. NaN is short for "not-a-number," and means that our calculation either didn't make mathematical sense or could not be performed properly. NA is short for "not available" and represents a missing value—a problem all too common in data analysis. In general, if our calculation involves a missing value, then the results will also be missing:

```
c(Inf + 1, Inf - 1, Inf - Inf)
## [1] Inf Inf NaN
c(1 / Inf, Inf / 1, Inf / Inf)
## [1]   0 Inf NaN
c(sqrt(Inf), sin(Inf))
## Warning: NaNs produced
## [1] Inf NaN
c(log(Inf), log(Inf, base = Inf))
```

---

2. Since the numbers are random, expect to get different values if you try this yourself.

```
## Warning: NaNs produced

## [1] Inf NaN

c(NA + 1, NA * 5, NA + Inf)

## [1] NA NA NA
```

When arithmetic involves NA and NaN, the answer is one of those two values, but which of those two is system dependent:

```
c(NA + NA, NaN + NaN, NaN + NA, NA + NaN)

## [1]  NA NaN NaN  NA
```

There are functions available to check for these special values. Notice that NaN and NA are neither finite nor infinite, and NaN is missing but NA *is* a number:

```
x <- c(0, Inf, -Inf, NaN, NA)
is.finite(x)

## [1]  TRUE FALSE FALSE FALSE FALSE

is.infinite(x)

## [1] FALSE  TRUE  TRUE FALSE FALSE

is.nan(x)

## [1] FALSE FALSE FALSE  TRUE FALSE

is.na(x)

## [1] FALSE FALSE FALSE  TRUE  TRUE
```

Logical Vectors

In addition to numbers, scientific calculation often involves logical values, particularly as a result of using the relational operators (<, etc.). Many programming languages use Boolean logic, where the values can be either TRUE or FALSE. In R, the situation is a little bit more complicated, since we can also have missing values, NA. This three-state system is sometimes call *troolean logic*, although that's a bad etymological joke, since the "Bool" in "Boolean" comes from George Bool, rather than anything to do with the word *binary*.

TRUE and FALSE are reserved words in R: you cannot create a variable with either of those names (lower- or mixed-case versions like True are fine, though). When you start R the variables T and F are already defined for you, taking the values TRUE and FALSE, respectively. This can save you a bit of typing, but it can also cause big problems. T and F are not reserved words, so users can redefine them. This means that it is OK to use the abbreviated names if you are tapping away at the command line, but not if your code is going to interact with someone else's (especially if their code involves Times or Temperatures or mathematical Functions).

There are three vectorized logical operators in R:

- ! is used for *not*.
- & is used for *and*.
- | is used for *or*.

```
(x <- 1:10 >= 5)
## [1] FALSE FALSE FALSE FALSE  TRUE  TRUE  TRUE  TRUE  TRUE  TRUE
!x
## [1]  TRUE  TRUE  TRUE  TRUE FALSE FALSE FALSE FALSE FALSE FALSE
(y <- 1:10 %% 2 == 0)
## [1] FALSE  TRUE FALSE  TRUE FALSE  TRUE FALSE  TRUE FALSE  TRUE
x & y
## [1] FALSE FALSE FALSE FALSE FALSE  TRUE FALSE  TRUE FALSE  TRUE
x | y
## [1] FALSE  TRUE FALSE  TRUE  TRUE  TRUE  TRUE  TRUE  TRUE  TRUE
```

We can conjure up some truth tables to see how they work (don't worry if this code doesn't make sense yet; just concentrate on understanding why each value occurs in the truth table):

```
x <- c(TRUE, FALSE, NA)          #the three logical values
xy <- expand.grid(x = x, y = x)  #get all combinations of x and y
within(                          #make the next assignments within xy
  xy,
  {
    and <- x & y
    or  <- x | y
    not.y <- !y
    not.x <- !x
  }
)
##       x     y not.x not.y    or   and
## 1  TRUE  TRUE FALSE FALSE  TRUE  TRUE
## 2 FALSE  TRUE  TRUE FALSE  TRUE FALSE
## 3    NA  TRUE    NA FALSE  TRUE    NA
## 4  TRUE FALSE FALSE  TRUE  TRUE FALSE
## 5 FALSE FALSE  TRUE  TRUE FALSE FALSE
## 6    NA FALSE    NA  TRUE    NA FALSE
## 7  TRUE    NA FALSE    NA  TRUE    NA
## 8 FALSE    NA  TRUE    NA    NA FALSE
## 9    NA    NA    NA    NA    NA    NA
```

Two other useful functions for dealing with logical vectors are any and all, which return TRUE if the input vector contains *at least* one TRUE value or *only* TRUE values, respectively:

```
none_true <- c(FALSE, FALSE, FALSE)
some_true <- c(FALSE, TRUE, FALSE)
all_true <- c(TRUE, TRUE, TRUE)
any(none_true)

## [1] FALSE

any(some_true)

## [1] TRUE

any(all_true)

## [1] TRUE

all(none_true)

## [1] FALSE

all(some_true)

## [1] FALSE

all(all_true)

## [1] TRUE
```

Summary

- R can be used as a very powerful scientific calculator.
- Assigning variables lets you reuse values.
- R has special values for positive and negative infinity, not-a-number, and missing values, to assist with mathematical operations.
- R uses troolean logic.

Test Your Knowledge: Quiz

Question 2-1
What is the operator used for integer division?

Question 2-2
How would you check if a variable, x, is equal to pi?

Question 2-3
Describe at least two ways of assigning a variable.

Question 2-4

Which of the five numbers 0, Inf, -Inf, NaN, and NA are infinite?

Question 2-5

Which of the five numbers 0, Inf, -Inf, NaN, and NA are considered not missing?

Test Your Knowledge: Exercises

Exercise 2-1

1. Calculate the inverse tangent (a.k.a. arctan) of the reciprocal of all integers from 1 to 1,000. Hint: take a look at the ?Trig help page to find the inverse tangent function. You don't need a function to calculate reciprocals. [5]

2. Assign the numbers 1 to 1,000 to a variable x. Calculate the inverse tangent of the reciprocal of x, as in part (a), and assign it to a variable y. Now reverse the operations by calculating the reciprocal of the tangent of y and assigning this value to a variable z. [5]

Exercise 2-2

Compare the variables x and z from Exercise 2-1 (b) using ==, identical, and all.equal. For all.equal, try changing the tolerance level by passing a third argument to the function. What happens if the tolerance is set to 0? [10]

Exercise 2-3

Define the following vectors:

1. true_and_missing, with the values TRUE and NA (at least one of each, in any order)

2. false_and_missing, with the values FALSE and NA

3. mixed, with the values TRUE, FALSE, and NA

Apply the functions any and all to each of your vectors. [5]

## CHAPTER 3

Inspecting Variables and Your Workspace

So far, we've run some calculations and assigned some variables. In this chapter, we'll find out ways to examine the properties of those variables and to manipulate the user workspace that contains them.

Chapter Goals

After reading this chapter, you should:

- Know what a class is, and the names of some common classes
- Know how to convert a variable from one type to another
- Be able to inspect variables to find useful information about them
- Be able to manipulate the user workspace

Classes

All variables in R have a class, which tells you what kinds of variables they are. For example, most numbers have class numeric (see the next section for the other types), and logical values have class logical. Actually, being picky about it, *vectors* of numbers are numeric and *vectors* of logical values are logical, since R has no scalar types. The "smallest" data type in R is a vector.

You can find out what the class of a variable is using *class(my_variable)*:

```
class(c(TRUE, FALSE))
## [1] "logical"
```

It's worth being aware that as well as a class, all variables also have an internal storage type (accessed via typeof), a mode (see mode), and a storage mode (storage.mode). If this sounds complicated, don't worry! Types, modes, and storage modes mostly exist for legacy purposes, so in practice you should only ever need to use an object's `class` (at least until you join the R Core Team). Appendix A has a reference table showing the relationships between class, type, and (storage) mode for many sorts of variables. Don't bother memorizing it, and don't worry if you don't recognize some of the classes. It is simply worth browsing the table to see which things are related to which other things.

From now on, to make things easier, I'm going to use "class" and "type" synonymously (except where noted).

Different Types of Numbers

All the variables that we created in the previous chapter were numbers, but R contains three different classes of numeric variable: `numeric` for floating point values; `integer` for, ahem, integers; and `complex` for complex numbers. We can tell which is which by examining the `class` of the variable:

```
class(sqrt(1:10))
## [1] "numeric"
class(3 + 1i)      #"i" creates imaginary components of complex numbers
## [1] "complex"
class(1)           #although this is a whole number, it has class numeric
## [1] "numeric"
class(1L)          #add a suffix of "L" to make the number an integer
## [1] "integer"
class(0.5:4.5)     #the colon operator returns a value that is numeric...
## [1] "numeric"
class(1:5)         #unless all its values are whole numbers
## [1] "integer"
```

Note that as of the time of writing, all floating point numbers are 32-bit numbers ("double precision"), even when installed on a 64-bit operating system, and 16-bit ("single precision") numbers don't exist.

Typing `.Machine` gives you some information about the properties of R's numbers. Although the values, in theory, can change from machine to machine, for most builds, most of the values are the same. Many of the values returned by `.Machine` need never concern you. It's worth knowing that the largest floating point number that R can represent at full precision is about `1.8e308`. This is big enough for everyday purposes, but a lot smaller than infinity! The smallest positive number that can be represented is

2.2e-308. Integers can take values up to 2 ^ 31 - 1, which is a little over two billion, (or down to -2 ^ 31 + 1).[1]

The only other value of much interest is ε, the smallest positive floating point number such that |ε + 1| != 1. That's a fancy way of saying how close two numbers can be so that R knows that they are different. It's about 2.2e-16. This value is used by all.equal when you compare two numeric vectors.

In fact, all of this is even easier than you think, since it is perfectly possible to get away with not (knowingly) using integers. R is designed so that anywhere an integer is needed —indexing a vector, for example—a floating point "numeric" number can be used just as well.

Other Common Classes

In addition to the three numeric classes and the logical class that we've seen already, there are three more classes of vectors: character for storing text, factors for storing categorical data, and the rarer raw for storing binary data.

In this next example, we create a character vector using the c operator, just like we did for numeric vectors. The class of a character vector is character:

```
class(c("she", "sells", "seashells", "on", "the", "sea", "shore"))
## [1] "character"
```

Note that unlike some languages, R doesn't distinguish between whole strings and individual characters—a string containing one character is treated the same as any other string. Unlike with some other lower-level languages, you don't need to worry about terminating your strings with a null character (\0). In fact, it is an error to try to include such a character in your strings.

In many programming languages, categorical data would be represented by integers. For example, gender could be represented as 1 for females and 2 for males. A slightly better solution would be to treat gender as a character variable with the choices "female" and "male." This is still semantically rather dubious, though, since categorical data is a different concept to plain old text. R has a more sophisticated solution that combines both these ideas in a semantically correct class—factors are integers with labels:

```
(gender <- factor(c("male", "female", "female", "male", "female")))
## [1] male    female female male    female
## Levels: female male
```

---

1. If these limits aren't good enough for you, higher-precision values are available via the Rmpfr package, and very large numbers are available in the brobdingnab package. These are fairly niche requirements, though; the three built-in classes of R numbers should be fine for almost all purposes.

The contents of the factor look much like their character equivalent—you get readable labels for each value. Those labels are confined to specific values (in this case "female" and "male") known as the *levels* of the factor:

```
levels(gender)

## [1] "female" "male"

nlevels(gender)

## [1] 2
```

Notice that even though "male" is the first value in gender, the first level is "female." By default, factor levels are assigned alphabetically.

Underneath the bonnet,[2] the factor values are stored as integers rather than characters. You can see this more clearly by calling as.integer:

```
as.integer(gender)

## [1] 2 1 1 2 1
```

This use of integers for storage makes them very memory-efficient compared to character text, at least when there are lots of repeated strings, as there are here. If we exaggerate the situation by generating 10,000 random genders (using the sample function to sample the strings "female" and "male" 10,000 times with replacement), we can see that a factor containing the values takes up less memory than the character equivalent. In the following code, sample returns a character vector—which we convert into a factor using as.factor--and object.size returns the memory allocation for each object:

```
gender_char <- sample(c("female", "male"), 10000, replace = TRUE)
gender_fac <- as.factor(gender_char)
object.size(gender_char)

## 80136 bytes

object.size(gender_fac)

## 40512 bytes
```

Variables take up different amounts of memory on 32-bit and 64-bit systems, so object.size will return different values in each case.

For manipulating the contents of factor levels (a common case would be cleaning up names, so all your men have the value "male" rather than "Male") it is typically best to

---

2. Or hood, if you prefer.

convert the factors to strings, in order to take advantage of string manipulation functions. You can do this in the obvious way, using as.character:

```
as.character(gender)
## [1] "male"    "female" "female" "male"    "female"
```

There is much more to learn about both character vectors and factors; they will be covered in depth in Chapter 7.

The raw class stores vectors of "raw" bytes.[3] Each byte is represented by a two-digit hexadecimal value. They are primarily used for storing the contents of imported binary files, and as such are reasonably rare. The integers 0 to 255 can be converted to raw using as.raw. Fractional and imaginary parts are discarded, and numbers outside this range are treated as 0. For strings, as.raw doesn't work; you must use charToRaw instead:

```
as.raw(1:17)
## [1] 01 02 03 04 05 06 07 08 09 0a 0b 0c 0d 0e 0f 10 11
as.raw(c(pi, 1 + 1i, -1, 256))
## Warning: imaginary parts discarded in coercion
## Warning: out-of-range values treated as 0 in coercion to raw
## [1] 03 01 00 00
(sushi <- charToRaw("Fish!"))
## [1] 46 69 73 68 21
class(sushi)
## [1] "raw"
```

As well as the vector classes that we've seen so far, there are many other types of variables; we'll spend the next few chapters looking at them.

Arrays contain multidimensional data, and matrices (via the matrix class) are the special case of two-dimensional arrays. They will be discussed in Chapter 4.

So far, all these variable types need to contain the same type of thing. For example, a character vector or array must contain all strings, and a logical vector or array must contain only logical values. Lists are flexible in that each item in them can be a different type, including other lists. Data frames are what happens when a matrix and a list have a baby. Like matrices, they are rectangular, and as in lists, each column can have a different type. They are ideal for storing spreadsheet-like data. Lists and data frames are discussed in Chapter 5.

---

3. It is unclear what a cooked byte would entail.

The preceding classes are all for storing data. Environments store the variables that store the data. As well as storing data, we clearly want to do things with it, and for that we need functions. We've already seen some functions, like sin and exp. In fact, operators like + are secretly functions too! Environments and functions will be discussed further in Chapter 6.

Chapter 7 discusses strings and factors in more detail, along with some options for storing dates and times.

There are some other types in R that are a little more complicated to understand, and we'll leave these until later. Formulae will be discussed in Chapter 15, and calls and expressions will be discussed in the section "Magic" on page 299 in Chapter 16. Classes will be discussed again in more depth in the section "Object-Oriented Programming" on page 302.

Checking and Changing Classes

Calling the class function is useful to interactively examine our variables at the command prompt, but if we want to test an object's type in our scripts, it is better to use the is function, or one of its class-specific variants. In a typical situation, our test will look something like:

```
if(!is(x, "some_class"))
{
  #some corrective measure
}
```

Most of the common classes have their own is.* functions, and calling these is usually a little bit more efficient than using the general is function. For example:

```
is.character("red lorry, yellow lorry")
```

```
## [1] TRUE
```

```
is.logical(FALSE)
```

```
## [1] TRUE
```

```
is.list(list(a = 1, b = 2))
```

```
## [1] TRUE
```

We can see a complete list of all the is functions in the base package using:

```
ls(pattern = "^is", baseenv())
```

```
##  [1] "is.array"        "is.atomic"
##  [3] "is.call"         "is.character"
##  [5] "is.complex"      "is.data.frame"
##  [7] "is.double"       "is.element"
##  [9] "is.environment"  "is.expression"
## [11] "is.factor"       "is.finite"
```

```
## [13] "is.function"             "is.infinite"
## [15] "is.integer"              "is.language"
## [17] "is.list"                 "is.loaded"
## [19] "is.logical"              "is.matrix"
## [21] "is.na"                   "is.na.data.frame"
## [23] "is.na.numeric_version"   "is.na.POSIXlt"
## [25] "is.na<-"                 "is.na<-.default"
## [27] "is.na<-.factor"          "is.name"
## [29] "is.nan"                  "is.null"
## [31] "is.numeric"              "is.numeric.Date"
## [33] "is.numeric.difftime"     "is.numeric.POSIXt"
## [35] "is.numeric_version"      "is.object"
## [37] "is.ordered"              "is.package_version"
## [39] "is.pairlist"             "is.primitive"
## [41] "is.qr"                   "is.R"
## [43] "is.raw"                  "is.recursive"
## [45] "is.single"               "is.symbol"
## [47] "is.table"                "is.unsorted"
## [49] "is.vector"               "isatty"
## [51] "isBaseNamespace"         "isdebugged"
## [53] "isIncomplete"            "isNamespace"
## [55] "isOpen"                  "isRestart"
## [57] "isS4"                    "isSeekable"
## [59] "isSymmetric"             "isSymmetric.matrix"
## [61] "isTRUE"
```

In the preceding example, ls lists variable names, "^is" is a regular expression that means "match strings that begin with 'is,'" and baseenv is a function that simply returns the environment of the base package. Don't worry what that means right now, since environments are quite an advanced topic; we'll return to them in Chapter 6.

The assertive package[4] contains more is functions with a consistent naming scheme.

One small oddity is that is.numeric returns TRUE for integers as well as floating point values. If we want to test for only floating point numbers, then we must use is.double. However, this isn't usually necessary, as R is designed so that floating point and integer values can be used more or less interchangeably. In the following examples, note that adding an L suffix makes the number into an integer:

```
is.numeric(1)
## [1] TRUE
is.numeric(1L)
## [1] TRUE
is.integer(1)
## [1] FALSE
```

---

4. Disclosure: I wrote it.

```
is.integer(1L)
```

```
## [1] TRUE
```

```
is.double(1)
```

```
## [1] TRUE
```

```
is.double(1L)
```

```
## [1] FALSE
```

Sometimes we may wish to change the type of an object. This is called *casting*, and most
`is*` functions have a corresponding `as*` function to achieve it. The specialized `as*`
functions should be used over plain `as` when available, since they are usually more
efficient, and often contain extra logic specific to that class. For example, when converting a string to a number, `as.numeric` is slightly more efficient than plain `as`, but
either can be used:

```
x <- "123.456"
as(x, "numeric")
```

```
## [1] 123.5
```

```
as.numeric(x)
```

```
## [1] 123.5
```

 The number of decimal places that R prints for numbers depends upon
your R setup. You can set a global default using `options(digits =n)`,
where *n* is between 1 and 22. Further control of printing numbers is
discussed in Chapter 7.

In this next example, however, note that when converting a vector into a data frame (a
variable for spreadsheet-like data), the general `as` function throws an error:

```
y <- c(2, 12, 343, 34997)        #See http://oeis.org/A192892
as(y, "data.frame")
as.data.frame(y)
```

 In general, the class-specific variants should always be used over standard `as`, if they are available.

It is also possible to change the type of an object by directly assigning it a new class,
though this isn't recommended (class assignment has a different use; see the section
"Object-Oriented Programming" on page 302):

```
x <- "123.456"
class(x) <- "numeric"
x
```

```
## [1] 123.5
```

```
is.numeric(x)
```

```
## [1] TRUE
```

Examining Variables

Whenever we've typed a calculation or the name of a variable at the console, the result has been printed. This happens because R implicitly calls the print method of the object.

 As a side note on terminology: "method" and "function" are basically interchangeable. Functions in R are sometimes called methods in an object-oriented context. There are different versions of the print function for different types of object, making matrices print differently from vectors, which is why I said "print method" here.

So, typing 1 + 1 at the command prompt does the same thing as print(1 + 1).

Inside loops or functions,[5] the automatic printing doesn't happen, so we have to explicitly call print:

```
ulams_spiral <- c(1, 8, 23, 46, 77)  #See http://oeis.org/A033951
for(i in ulams_spiral) i             #uh-oh, the values aren't printed
for(i in ulams_spiral) print(i)
```

```
## [1] 1
## [1] 8
## [1] 23
## [1] 46
## [1] 77
```

This is also true on some systems if you run R from a terminal rather than using a GUI or IDE. In this case you will always need to explicitly call the print function.

Most print functions are built upon calls to the lower-level cat function. You should almost never have to call cat directly (print and message are the user-level equivalents), but it is worth knowing in case you ever need to write your own print function.[6]

---

5. Except for the value being returned from the function.

6. Like in Exercise 16-3, perhaps.

Both the c and cat functions are short for *concatenate*, though they perform quite different tasks! cat is named after a Unix function.

As well as viewing the printout of a variable, it is often helpful to see some sort of summary of the object. The summary function does just that, giving appropriate information for different data types. Numeric variables are summarized as mean, median, and some quantiles. Here, the runif function generates 30 random numbers that are uniformly distributed between 0 and 1:

```
num <- runif(30)
summary(num)

##    Min. 1st Qu.  Median    Mean 3rd Qu.    Max.
##  0.0211  0.2960  0.5060  0.5290  0.7810  0.9920
```

Categorical and logical vectors are summarized by the counts of each value. In this next example, letters is a built-in constant that contains the lowercase values from "a" to "z" (LETTERS contains the uppercase equivalents, "A" to "Z"). Here, letters[1:5] uses indexing to restrict the letters to "a" to "e." The sample function randomly samples these values, with replace, 30 times:

```
fac <- factor(sample(letters[1:5], 30, replace = TRUE))
summary(fac)

## a b c d e
## 6 7 5 9 3

bool <- sample(c(TRUE, FALSE, NA), 30, replace = TRUE)
summary(bool)

##    Mode   FALSE    TRUE    NA's
## logical      12      11       7
```

Multidimensional objects, like matrices and data frames, are summarized by column (we'll look at these in more detail in the next two chapters). The data frame dfr that we create here is quite large to display, having 30 rows. For large objects like this,[7] the head function can be used to display only the first few rows (six by default):

```
dfr <- data.frame(num, fac, bool)
head(dfr)

##        num fac  bool
## 1 0.47316   b    NA
## 2 0.56782   d FALSE
## 3 0.46205   d FALSE
## 4 0.02114   b  TRUE
```

7. These days, 30 rows isn't usually considered to be "big data," but it's still a screenful when printed.

```
## 5 0.27963   a   TRUE
## 6 0.46690   a   TRUE
```

The summary function for data frames works like calling summary on each individual column:

```
summary(dfr)
```

```
##       num         fac      bool
## Min.   :0.0211   a:6   Mode :logical
## 1st Qu.:0.2958   b:7   FALSE:12
## Median :0.5061   c:5   TRUE :11
## Mean   :0.5285   d:9   NA's :7
## 3rd Qu.:0.7808   e:3
## Max.   :0.9916
```

Similarly, the str function shows the object's structure. It isn't that interesting for vectors (since they are so simple), but str is exceedingly useful for data frames and nested lists:

```
str(num)
```

```
## num [1:30] 0.4732 0.5678 0.462 0.0211 0.2796 ...
```

```
str(dfr)
```

```
## 'data.frame':    30 obs. of  3 variables:
## $ num : num  0.4732 0.5678 0.462 0.0211 0.2796 ...
## $ fac : Factor w/ 5 levels "a","b","c","d",..: 2 4 4 2 1 1 4 2 1 4 ...
## $ bool: logi  NA FALSE FALSE TRUE TRUE TRUE ...
```

As mentioned previously, each class typically has its own print method that controls how it is displayed to the console. Sometimes this printing obscures its internal structure, or omits useful information. The unclass function can be used to bypass this, letting you see how a variable is constructed. For example, calling unclass on a factor reveals that it is just an integer vector, with an attribute called levels:

```
unclass(fac)
```

```
## [1] 2 4 4 2 1 1 4 2 1 4 3 3 1 5 4 5 1 5 1 2 2 3 4 2 4 3 4 2 3 4
## attr(,"levels")
## [1] "a" "b" "c" "d" "e"
```

We'll look into attributes later on, but for now, it is useful to know that the attributes function gives you a list of all the attributes belonging to an object:

```
attributes(fac)
```

```
## $levels
## [1] "a" "b" "c" "d" "e"
##
## $class
## [1] "factor"
```

For visualizing two-dimensional variables such as matrices and data frames, the View function (notice the capital "V") displays the variable as a (read-only) spreadsheet. The edit and fix functions work similarly to View, but let us manually change data values. While this may sound more useful, it is usually a supremely awful idea to edit data in this way, since we lose all traceability of where our data came from. It is almost always better to edit data programmatically:

```
View(dfr)              #no changes allowed
new_dfr <- edit(dfr)   #changes stored in new_dfr
fix(dfr)               #changes stored in dfr
```

A useful trick is to view the first few rows of a data frame by combining View with head:

```
View(head(dfr, 50))    #view first 50 rows
```

The Workspace

While we're working, it's often nice to know the names of the variables that we've created and what they contain. To list the names of existing variables, use the function ls. This is named after the equivalent Unix command, and follows the same convention: by default, variable names that begin with a . are hidden. To see them, pass the argument all.names = TRUE:

```
#Create some variables to find
peach <- 1
plum <- "fruity"
pear <- TRUE
ls()
```

```
##  [1] "a_vector"        "all_true"       "bool"
##  [4] "dfr"             "fac"            "fname"
##  [7] "gender"          "gender_char"    "gender_fac"
## [10] "i"               "input"          "my_local_variable"
## [13] "none_true"       "num"            "output"
## [16] "peach"           "pear"           "plum"
## [19] "remove_package"  "some_true"      "sushi"
## [22] "ulams_spiral"    "x"              "xy"
## [25] "y"               "z"              "zz"
```

```
ls(pattern = "ea")
```

```
## [1] "peach" "pear"
```

For more information about our workspace, we can see the structure of our variables using ls.str. This is, as you might expect, a combination of the ls and str functions, and is very useful during debugging sessions (see "Debugging" on page 292 in Chapter 16). browseEnv provides a similar capability, but displays its output in an HTML page in our web browser:

```
browseEnv()
```

After working for a while, especially while exploring data, our workspace can become quite cluttered. We can clean it up by using the rm function to remove variables:

```
rm(peach, plum, pear)
rm(list = ls())        #Removes everything. Use with caution!
```

Summary

- All variables have a *class*.
- You test if an object has a particular class using the is function, or one of its class-specific variants.
- You can change the class of an object using the as function, or one of its class-specific variants.
- There are several functions that let you inspect variables, including summary, head, str, unclass, attributes, and View.
- ls lists the names of your variables and ls.str lists them along with their structure.
- rm removes your variables.

Test Your Knowledge: Quiz

Question 3-1
    What are the names of the three built-in classes of numbers?

Question 3-2
    What function would you call to find out the number of levels of a factor?

Question 3-3
    How might you convert the string "6.283185" to a number?

Question 3-4
    Name at least three functions for inspecting the contents of a variable.

Question 3-5
    How would you remove all the variables in the user workspace?

Test Your Knowledge: Exercises

Exercise 3-1
    Find the class, type, mode, and storage mode of the following values: Inf, NA, NaN, "". [5]

Exercise 3-2

Randomly generate 1,000 pets, from the choices "dog," "cat," "hamster," and "gold-fish," with equal probability of each being chosen. Display the first few values of the resultant variable, and count the number of each type of pet. [5]

Exercise 3-3

Create some variables named after vegetables. List the names of all the variables in the user workspace that contain the letter "a." [5]

Vectors, Matrices, and Arrays

In Chapters 1 and 2, we saw several types of vectors for logical values, character strings, and of course numbers. This chapter shows you more manipulation techniques for vectors and introduces their multidimensional brethren, matrices and arrays.

Chapter Goals

After reading this chapter, you should:

- Be able to create new vectors from existing vectors
- Understand lengths, dimensions, and names
- Be able to create and manipulate matrices and arrays

Vectors

So far, you have used the colon operator, :, for creating sequences from one number to another, and the c function for concatenating values and vectors to create longer vectors. To recap:

```
8.5:4.5              #sequence of numbers from 8.5 down to 4.5
## [1] 8.5 7.5 6.5 5.5 4.5
c(1, 1:3, c(5, 8), 13) #values concatenated into single vector
## [1]  1  1  2  3  5  8 13
```

The vector function creates a vector of a specified type and length. Each of the values in the result is zero, FALSE, or an empty string, or whatever the equivalent of "nothing" is:

```
vector("numeric", 5)
## [1] 0 0 0 0 0
vector("complex", 5)
## [1] 0+0i 0+0i 0+0i 0+0i 0+0i
vector("logical", 5)
## [1] FALSE FALSE FALSE FALSE FALSE
vector("character", 5)
## [1] "" "" "" "" ""
vector("list", 5)
## [[1]]
## NULL
##
## [[2]]
## NULL
##
## [[3]]
## NULL
##
## [[4]]
## NULL
##
## [[5]]
## NULL
```

In that last example, NULL is a special "empty" value (not to be confused with NA, which indicates a missing data point). We'll look at NULL in detail in Chapter 5. For convenience, wrapper functions exist for each type to save you typing when creating vectors in this way. The following commands are equivalent to the previous ones:

```
numeric(5)
## [1] 0 0 0 0 0
complex(5)
## [1] 0+0i 0+0i 0+0i 0+0i 0+0i
logical(5)
## [1] FALSE FALSE FALSE FALSE FALSE
character(5)
## [1] "" "" "" "" ""
```

 As we'll see in the next chapter, the list function does not work the same way. list(5) creates something a little different.

Sequences

Beyond the colon operator, there are several functions for creating more general sequences. The seq function is the most general, and allows you to specify sequences in many different ways. In practice, though, you should never need to call it, since there are three other specialist sequence functions that are faster and easier to use, covering specific use cases.

seq.int lets us create a sequence from one number to another. With two inputs, it works exactly like the colon operator:

```
seq.int(3, 12)      #same as 3:12
## [1]  3  4  5  6  7  8  9 10 11 12
```

seq.int is slightly more general than :, since it lets you specify how far apart intermediate values should be:

```
seq.int(3, 12, 2)
## [1]  3  5  7  9 11
seq.int(0.1, 0.01, -0.01)
##  [1] 0.10 0.09 0.08 0.07 0.06 0.05 0.04 0.03 0.02 0.01
```

seq_len creates a sequence from 1 up to its input, so seq_len(5) is just a clunkier way of writing 1:5. However, the function is extremely useful for situations when its input could be zero:

```
n <- 0
1:n         #not what you might expect!
## [1] 1 0
seq_len(n)
## integer(0)
```

seq_along creates a sequence from 1 up to the length of its input:

```
pp <- c("Peter", "Piper", "picked", "a", "peck", "of", "pickled", "peppers")
for(i in seq_along(pp)) print(pp[i])
## [1] "Peter"
## [1] "Piper"
## [1] "picked"
## [1] "a"
## [1] "peck"
## [1] "of"
## [1] "pickled"
## [1] "peppers"
```

For each of the preceding examples, you can replace seq.int, seq_len, or seq_along with plain seq and get the same answer, though there is no need to do so.

Lengths

I've just sneakily introduced a new concept related to vectors. That is, all vectors have a *length*, which tells us how many elements they contain. This is a nonnegative integer[1] (yes, zero-length vectors are allowed), and you can access this value with the length function. Missing values still count toward the length:

```
length(1:5)
```

```
## [1] 5
```

```
length(c(TRUE, FALSE, NA))
```

```
## [1] 3
```

One possible source of confusion is character vectors. With these, the length is the number of strings, not the number of characters in each string. For that, we should use nchar:

```
sn <- c("Sheena", "leads", "Sheila", "needs")
length(sn)
```

```
## [1] 4
```

```
nchar(sn)
```

```
## [1] 6 5 6 5
```

It is also possible to assign a new length to a vector, but this is an unusual thing to do, and probably indicates bad code. If you shorten a vector, the values at the end will be removed, and if you extend a vector, missing values will be added to the end:

```
poincare <- c(1, 0, 0, 0, 2, 0, 2, 0)  #See http://oeis.org/A051629
length(poincare) <- 3
poincare
```

```
## [1] 1 0 0
```

```
length(poincare) <- 8
poincare
```

```
## [1]  1  0  0 NA NA NA NA NA
```

Names

A great feature of R's vectors is that each element can be given a name. Labeling the elements can often make your code much more readable. You can specify names when you create a vector in the form *name = value*. If the name of an element is a valid variable name, it doesn't need to be enclosed in quotes. You can name some elements of a vector and leave others blank:

---

1. Lengths are limited to 2^31-1 elements on 32-bit systems and versions of R prior to 3.0.0.

```
c(apple = 1, banana = 2, "kiwi fruit" = 3, 4)
```

```
##     apple    banana kiwi fruit
##         1         2          3          4
```

You can add element names to a vector after its creation using the names function:

```
x <- 1:4
names(x) <- c("apple", "bananas", "kiwi fruit", "")
x
```

```
##     apple   bananas kiwi fruit
##         1         2          3          4
```

This names function can also be used to retrieve the names of a vector:

```
names(x)
```

```
## [1] "apple"      "bananas"    "kiwi fruit" ""
```

If a vector has no element names, then the names function returns NULL:

```
names(1:4)
```

```
## NULL
```

Indexing Vectors

Oftentimes we may want to access only part of a vector, or perhaps an individual element. This is called *indexing* and is accomplished with square brackets, [ ]. (Some people also call it *subsetting* or *subscripting* or *slicing*. All these terms refer to the same thing.) R has a very flexible system that gives us several choices of index:

- Passing a vector of positive numbers returns the slice of the vector containing the elements at those locations. The first position is 1 (not 0, as in some other languages).
- Passing a vector of negative numbers returns the slice of the vector containing the elements everywhere *except* at those locations.
- Passing a logical vector returns the slice of the vector containing the elements where the index is TRUE.
- For named vectors, passing a character vector of names returns the slice of the vector containing the elements with those names.

Consider this vector:

```
x <- (1:5) ^ 2
```

```
## [1]  1  4  9 16 25
```

These three indexing methods return the same values:

```
x[c(1, 3, 5)]
```

```
x[c(-2, -4)]
```

```
x[c(TRUE, FALSE, TRUE, FALSE, TRUE)]
```

```
## [1]  1  9 25
```

After naming each element, this method also returns the same values:

```
names(x) <- c("one", "four", "nine", "sixteen", "twenty five")
x[c("one", "nine", "twenty five")]
```

```
##          one        nine twenty five
##            1           9          25
```

Mixing positive and negative values is not allowed, and will throw an error:

```
x[c(1, -1)]      #This doesn't make sense!
```

```
## Error: only 0's may be mixed with negative subscripts
```

If you use positive numbers or logical values as the index, then missing indices correspond to missing values in the result:

```
x[c(1, NA, 5)]
```

```
##          one        <NA> twenty five
##            1          NA          25
```

```
x[c(TRUE, FALSE, NA, FALSE, TRUE)]
```

```
##          one        <NA> twenty five
##            1          NA          25
```

Missing values don't make any sense for negative indices, and cause an error:

```
x[c(-2, NA)]      #This doesn't make sense either!
```

```
## Error: only 0's may be mixed with negative subscripts
```

Out of range indices, beyond the length of the vector, don't cause an error, but instead return the missing value NA. In practice, it is usually better to make sure that your indices are in range than to use out of range values:

```
x[6]
```

```
## <NA>
##   NA
```

Noninteger indices are silently rounded toward zero. This is another case where R is arguably too permissive. If you find yourself passing fractions as indices, you are probably writing bad code:

```
x[1.9]   #1.9 rounded to 1
```

```
## one
##   1
```

```
x[-1.9]  #-1.9 rounded to -1
```

```
##         four        nine     sixteen twenty five
##            4           9          16          25
```

Not passing any index will return the whole of the vector, but again, if you find yourself not passing any index, then you are probably doing something odd:

```
x[]
##         one        four        nine      sixteen twenty five
##           1           4           9           16           25
```

The `which` function returns the locations where a logical vector is `TRUE`. This can be useful for switching from logical indexing to integer indexing:

```
which(x > 10)
##      sixteen twenty five
##           4           5
```

`which.min` and `which.max` are more efficient shortcuts for `which(min(x))` and `which(max(x))`, respectively:

```
which.min(x)
## one
##   1

which.max(x)
## twenty five
##           5
```

Vector Recycling and Repetition

So far, all the vectors that we have added together have been the same length. You may be wondering, "What happens if I try to do arithmetic on vectors of different lengths?"

If we try to add a single number to a vector, then that number is added to each element of the vector:

```
1:5 + 1
## [1] 2 3 4 5 6

1 + 1:5
## [1] 2 3 4 5 6
```

When adding two vectors together, R will recycle elements in the shorter vector to match the longer one:

```
1:5 + 1:15
## [1]  2  4  6  8 10  7  9 11 13 15 12 14 16 18 20
```

If the length of the longer vector isn't a multiple of the length of the shorter one, a warning will be given:

```
1:5 + 1:7
## Warning: longer object length is not a multiple of shorter object length
## [1]  2  4  6  8 10  7  9
```

It must be stressed that just because we *can* do arithmetic on vectors of different lengths, it doesn't mean that we *should*. Adding a scalar value to a vector is okay, but otherwise we are liable to get ourselves confused. It is much better to explicitly create equal-length vectors before we operate on them.

The `rep` function is very useful for this task, letting us create a vector with repeated elements:

```
rep(1:5, 3)
## [1] 1 2 3 4 5 1 2 3 4 5 1 2 3 4 5
rep(1:5, each = 3)
## [1] 1 1 1 2 2 2 3 3 3 4 4 4 5 5 5
rep(1:5, times = 1:5)
## [1] 1 2 2 3 3 3 4 4 4 4 5 5 5 5 5
rep(1:5, length.out = 7)
## [1] 1 2 3 4 5 1 2
```

Like the `seq` function, `rep` has a simpler and faster variant, `rep.int`, for the most common case:

```
rep.int(1:5, 3)   #the same as rep(1:5, 3)
## [1] 1 2 3 4 5 1 2 3 4 5 1 2 3 4 5
```

Recent versions of R (since v3.0.0) also have `rep_len`, paralleling `seq_len`, which lets us specify the length of the output vector:

```
rep_len(1:5, 13)
## [1] 1 2 3 4 5 1 2 3 4 5 1 2 3
```

Matrices and Arrays

The vector variables that we have looked at so far are one-dimensional objects, since they have length but no other dimensions. Arrays hold multidimensional rectangular data. "Rectangular" means that each row is the same length, and likewise for each column and other dimensions. Matrices are a special case of two-dimensional arrays.

Creating Arrays and Matrices

To create an array, you call the `array` function, passing in a vector of values and a vector of dimensions. Optionally, you can also provide names for each dimension:

```
(three_d_array <- array(
  1:24,
  dim = c(4, 3, 2),
  dimnames = list(
    c("one", "two", "three", "four"),
    c("ein", "zwei", "drei"),
    c("un", "deux")
  )
))
## , , un
##
##       ein zwei drei
## one     1    5    9
## two     2    6   10
## three   3    7   11
## four    4    8   12
##
## , , deux
##
##       ein zwei drei
## one    13   17   21
## two    14   18   22
## three  15   19   23
## four   16   20   24

class(three_d_array)

## [1] "array"
```

The syntax for creating matrices is similar, but rather than passing a `dim` argument, you specify the number of rows or the number of columns:

```
(a_matrix <- matrix(
  1:12,
  nrow = 4,              #ncol = 3 works the same
  dimnames = list(
    c("one", "two", "three", "four"),
    c("ein", "zwei", "drei")
  )
))
##       ein zwei drei
## one     1    5    9
## two     2    6   10
## three   3    7   11
## four    4    8   12

class(a_matrix)

## [1] "matrix"
```

This matrix could also be created using the `array` function. The following two-dimensional array is identical to the matrix that we just created (it even has class `matrix`):

```
(two_d_array <- array(
  1:12,
  dim = c(4, 3),
  dimnames = list(
    c("one", "two", "three", "four"),
    c("ein", "zwei", "drei")
  )
))
##       ein zwei drei
## one     1    5    9
## two     2    6   10
## three   3    7   11
## four    4    8   12

identical(two_d_array, a_matrix)
## [1] TRUE

class(two_d_array)
## [1] "matrix"
```

When you create a matrix, the values that you passed in fill the matrix column-wise. It is also possible to fill the matrix row-wise by specifying the argument byrow = TRUE:

```
matrix(
  1:12,
  nrow = 4,
  byrow = TRUE,
  dimnames = list(
    c("one", "two", "three", "four"),
    c("ein", "zwei", "drei")
  )
)
##       ein zwei drei
## one     1    2    3
## two     4    5    6
## three   7    8    9
## four   10   11   12
```

Rows, Columns, and Dimensions

For both matrices and arrays, the dim function returns a vector of integers of the dimensions of the variable:

```
dim(three_d_array)
## [1] 4 3 2

dim(a_matrix)
## [1] 4 3
```

For matrices, the functions `nrow` and `ncol` return the number of rows and columns, respectively:

```
nrow(a_matrix)
## [1] 4
ncol(a_matrix)
## [1] 3
```

`nrow` and `ncol` also work on arrays, returning the first and second dimensions, respectively, but it is usually better to use `dim` for higher-dimensional objects:

```
nrow(three_d_array)
## [1] 4
ncol(three_d_array)
## [1] 3
```

The `length` function that we have previously used with vectors also works on matrices and arrays. In this case it returns the product of each of the dimensions:

```
length(three_d_array)
## [1] 24
length(a_matrix)
## [1] 12
```

We can also reshape a matrix or array by assigning a new dimension with `dim`. This should be used with caution since it strips dimension names:

```
dim(a_matrix) <- c(6, 2)
a_matrix
##      [,1] [,2]
## [1,]    1    7
## [2,]    2    8
## [3,]    3    9
## [4,]    4   10
## [5,]    5   11
## [6,]    6   12
```

`nrow`, `ncol`, and `dim` return `NULL` when applied to vectors. The functions `NROW` and `NCOL` are counterparts to `nrow` and `ncol` that pretend vectors are matrices with a single column (that is, column vectors in the mathematical sense):

```
identical(nrow(a_matrix), NROW(a_matrix))
## [1] TRUE
identical(ncol(a_matrix), NCOL(a_matrix))
## [1] TRUE
```

```
recaman <- c(0, 1, 3, 6, 2, 7, 13, 20)
nrow(recaman)
## NULL
NROW(recaman)
## [1] 8
ncol(recaman)
## NULL
NCOL(recaman)
## [1] 1
dim(recaman)
```

Row, Column, and Dimension Names

In the same way that vectors have names for the elements, matrices have rownames and colnames for the rows and columns. For historical reasons, there is also a function row.names, which does the same thing as rownames, but there is no corresponding col.names, so it is better to ignore it and use rownames instead. As with the case of nrow, ncol, and dim, the equivalent function for arrays is dimnames. The latter returns a list (see "Lists" on page 57) of character vectors. In the following code chunk, a_matrix has been restored to its previous state, before its dimensions were changed:

```
rownames(a_matrix)

## [1] "one"    "two"    "three" "four"

colnames(a_matrix)

## [1] "ein"  "zwei" "drei"

dimnames(a_matrix)

## [[1]]
## [1] "one"    "two"    "three" "four"
##
## [[2]]
## [1] "ein"  "zwei" "drei"

rownames(three_d_array)

## [1] "one"    "two"    "three" "four"

colnames(three_d_array)

## [1] "ein"  "zwei" "drei"

dimnames(three_d_array)
```

```
## [[1]]
## [1] "one"    "two"    "three" "four"
##
## [[2]]
## [1] "ein"   "zwei" "drei"
##
## [[3]]
## [1] "un"    "deux"
```

Indexing Arrays

Indexing works just like it does with vectors, except that now we have to specify an index for more than one dimension. As before, we use square brackets to denote an index, and we still have four choices for specifying the index (positive integers, negative integers, logical values, and element names). It is perfectly permissible to specify the indices for different dimensions in different ways. The indices for each dimension are separated by commas:

```
a_matrix[1, c("zwei", "drei")] #elements in 1st row, 2nd and 3rd columns

## zwei drei
##    5    9
```

To include all of a dimension, leave the corresponding index blank:

```
a_matrix[1, ]                    #all of the first row

##  ein zwei drei
##    1    5    9

a_matrix[, c("zwei", "drei")]   #all of the second and third columns

##       zwei drei
## one      5    9
## two      6   10
## three    7   11
## four     8   12
```

Combining Matrices

The c function converts matrices to vectors before concatenating them:

```
(another_matrix <- matrix(
  seq.int(2, 24, 2),
  nrow = 4,
  dimnames = list(
    c("five", "six", "seven", "eight"),
    c("vier", "funf", "sechs")
  )
))
```

```
##        vier funf sechs
## five      2   10    18
## six       4   12    20
## seven     6   14    22
## eight     8   16    24

c(a_matrix, another_matrix)

##  [1]  1  2  3  4  5  6  7  8  9 10 11 12  2  4  6  8 10 12 14 16 18 20 22
## [24] 24
```

More natural combining of matrices can be achieved by using `cbind` and `rbind`, which
bind matrices together by columns and rows:

```
cbind(a_matrix, another_matrix)

##        ein zwei drei vier funf sechs
## one      1    5    9    2   10    18
## two      2    6   10    4   12    20
## three    3    7   11    6   14    22
## four     4    8   12    8   16    24

rbind(a_matrix, another_matrix)

##        ein zwei drei
## one      1    5    9
## two      2    6   10
## three    3    7   11
## four     4    8   12
## five     2   10   18
## six      4   12   20
## seven    6   14   22
## eight    8   16   24
```

Array Arithmetic

The standard arithmetic operators (+, -, *, /) work element-wise on matrices and
arrays, just they like they do on vectors:

```
a_matrix + another_matrix

##        ein zwei drei
## one      3   15   27
## two      6   18   30
## three    9   21   33
## four    12   24   36

a_matrix * another_matrix

##        ein zwei drei
## one      2   50  162
## two      8   72  200
## three   18   98  242
## four    32  128  288
```

When performing arithmetic on two arrays, you need to make sure that they are of an appropriate size (they must be "conformable," in linear algebra terminology). For example, both arrays must be the same size when adding, and for multiplication the number of rows in the first matrix must be the same as the number of columns in the second matrix:

```
(another_matrix <- matrix(1:12, nrow = 2))
a_matrix + another_matrix   #adding nonconformable matrices throws an error
```

If you try to add a vector to an array, then the usual vector recycling rules apply, but the dimension of the results is taken from the array.

The t function transposes matrices (but not higher-dimensional arrays, where the concept isn't well defined):

```
t(a_matrix)
```

```
##      one two three four
## ein    1   2     3    4
## zwei   5   6     7    8
## drei   9  10    11   12
```

For inner and outer matrix multiplication, we have the special operators %*% and %o%. In each case, the dimension names are taken from the first input, if they exist:

```
a_matrix %*% t(a_matrix)   #inner multiplication
```

```
##       one two three four
## one   107 122   137  152
## two   122 140   158  176
## three 137 158   179  200
## four  152 176   200  224
```

```
1:3 %o% 4:6                 #outer multiplication
```

```
##      [,1] [,2] [,3]
## [1,]    4    5    6
## [2,]    8   10   12
## [3,]   12   15   18
```

```
outer(1:3, 4:6)             #same
```

```
##      [,1] [,2] [,3]
## [1,]    4    5    6
## [2,]    8   10   12
## [3,]   12   15   18
```

The power operator, ^, also works element-wise on matrices, so to invert a matrix you cannot simply raise it to the power of minus one. Instead, this can be done using the solve function:[2]

---

2. qr.solve(m) and chol2inv(chol(m)) provide alternative algorithms for inverting matrices, but solve should be your first port of call.

```
(m <- matrix(c(1, 0, 1, 5, -3, 1, 2, 4, 7), nrow = 3))
```

```
##      [,1] [,2] [,3]
## [1,]   1    5    2
## [2,]   0   -3    4
## [3,]   1    1    7
```

```
m ^ -1
```

```
##      [,1]    [,2]   [,3]
## [1,]   1  0.2000 0.5000
## [2,] Inf -0.3333 0.2500
## [3,]   1  1.0000 0.1429
```

```
(inverse_of_m <- solve(m))
```

```
##      [,1] [,2] [,3]
## [1,]  -25  -33   26
## [2,]    4    5   -4
## [3,]    3    4   -3
```

```
m %*% inverse_of_m
```

```
##      [,1] [,2] [,3]
## [1,]    1    0    0
## [2,]    0    1    0
## [3,]    0    0    1
```

Summary

- seq and its variants let you create sequences of numbers.
- Vectors have a length that can be accessed or set with the length function.
- You can name elements of vectors, either when they are created or with the names function.
- You can access slices of a vector by passing an index into square brackets. The rep function creates a vector with repeated elements.
- Arrays are multidimensional objects, with matrices being the special case of two-dimensional arrays.
- nrow, ncol, and dim provide ways of accessing the dimensions of an array.
- Likewise, rownames, colnames, and dimnames access the names of array dimensions.

Test Your Knowledge: Quiz

Question 4-1
How would you create a vector containing the values 0, 0.25, 0.5, 0.75, and 1?

Question 4-2
Describe two ways of naming elements in a vector.

Question 4-3
What are the four types of index for a vector?

Question 4-4
What is the length of a 3-by-4-by-5 array?

Question 4-5
Which operator would you use to perform an inner product on two matrices?

Test Your Knowledge: Exercises

Exercise 4-1
The nth triangular number is given by `n * (n + 1) / 2`. Create a sequence of the first 20 triangular numbers. R has a built-in constant, `letters`, that contains the lowercase letters of the Roman alphabet. Name the elements of the vector that you just created with the first 20 letters of the alphabet. Select the triangular numbers where the name is a vowel. [10]

Exercise 4-2
The `diag` function has several uses, one of which is to take a vector as its input and create a square matrix with that vector on the diagonal. Create a 21-by-21 matrix with the sequence 10 to 0 to 11 (i.e., 11, 10, … , 1, 0, 1, …, 11). [5]

Exercise 4-3
By passing two extra arguments to `diag`, you can specify the dimensions of the output. Create a 20-by-21 matrix with ones on the main diagonal. Now add a row of zeros above this to create a 21-by-21 square matrix, where the ones are offset a row below the main diagonal.

Create another matrix with the ones offset one up from the diagonal.

Add these two matrices together, then add the answer from Exercise 4-2. The resultant matrix is called a Wilkinson matrix.

The `eigen` function calculates eigenvalues and eigenvectors of a matrix. Calculate the eigenvalues for your Wilkinson matrix. What do you notice about them? [20]

## CHAPTER 5

Lists and Data Frames

The vectors, matrices, and arrays that we have seen so far contain elements that are all of the same type. Lists and data frames are two types that let us combine different types of data in a single variable.

Chapter Goals

After reading this chapter, you should:

- Be able to create lists and data frames
- Be able to use `length`, `names`, and other functions to inspect and manipulate these types
- Understand what `NULL` is and when to use it
- Understand the difference between recursive and atomic variables
- Know how to perform basic manipulation of lists and data frames

Lists

A list is, loosely speaking, a vector where each element can be of a different type. This section concerns how to create, index, and manipulate lists.

Creating Lists

Lists are created with the `list` function, and specifying the contents works much like the c function that we've seen already. You simply list the contents, with each argument separated by a comma. List elements can be any variable type—vectors, matrices, even functions:

```
(a_list <- list(
  c(1, 1, 2, 5, 14, 42),    #See http://oeis.org/A000108
  month.abb,
  matrix(c(3, -8, 1, -3), nrow = 2),
  asin
))
## [[1]]
## [1]  1  1  2  5 14 42
##
## [[2]]
##  [1] "Jan" "Feb" "Mar" "Apr" "May" "Jun" "Jul" "Aug" "Sep" "Oct" "Nov"
## [12] "Dec"
##
## [[3]]
##      [,1] [,2]
## [1,]    3    1
## [2,]   -8   -3
##
## [[4]]
## function (x)  .Primitive("asin")
```

As with vectors, you can name elements during construction, or afterward using the names function:

```
names(a_list) <- c("catalan", "months", "involutary", "arcsin")
a_list

## $catalan
## [1]  1  1  2  5 14 42
##
## $months
##  [1] "Jan" "Feb" "Mar" "Apr" "May" "Jun" "Jul" "Aug" "Sep" "Oct" "Nov"
## [12] "Dec"
##
## $involutary
##      [,1] [,2]
## [1,]    3    1
## [2,]   -8   -3
##
## $arcsin
## function (x)  .Primitive("asin")

(the_same_list <- list(
  catalan    = c(1, 1, 2, 5, 14, 42),
  months     = month.abb,
  involutary = matrix(c(3, -8, 1, -3), nrow = 2),
  arcsin     = asin
))

## $catalan
## [1]  1  1  2  5 14 42
##
## $months
```

```
## [1] "Jan" "Feb" "Mar" "Apr" "May" "Jun" "Jul" "Aug" "Sep" "Oct" "Nov"
## [12] "Dec"
##
## $involutary
##      [,1] [,2]
## [1,]   3    1
## [2,]  -8   -3
##
## $arcsin
## function (x)  .Primitive("asin")
```

It isn't compulsory, but it helps if the names that you give elements are valid variable names.

It is even possible for elements of lists to be lists themselves:

```
(main_list <- list(
  middle_list             = list(
    element_in_middle_list = diag(3),
    inner_list             = list(
      element_in_inner_list         = pi ^ 1:4,
      another_element_in_inner_list = "a"
    )
  ),
  element_in_main_list = log10(1:10)
))
## $middle_list
## $middle_list$element_in_middle_list
##      [,1] [,2] [,3]
## [1,]   1    0    0
## [2,]   0    1    0
## [3,]   0    0    1
##
## $middle_list$inner_list
## $middle_list$inner_list$element_in_inner_list
## [1] 3.142
##
## $middle_list$inner_list$another_element_in_inner_list
## [1] "a"
##
##
##
## $element_in_main_list
##  [1] 0.0000 0.3010 0.4771 0.6021 0.6990 0.7782 0.8451 0.9031 0.9542 1.0000
```

In theory, you can keep nesting lists forever. In practice, current versions of R will throw an error once you start nesting your lists tens of thousands of levels deep (the exact number is machine specific). Luckily, this shouldn't be a problem for you, since real-world code where nesting is deeper than three or four levels is extremely rare.

Atomic and Recursive Variables

Due to this ability to contain other lists within themselves, lists are considered to be *recursive* variables. Vectors, matrices, and arrays, by contrast, are *atomic*. (Variables can either be recursive or atomic, never both; Appendix A contains a table explaining which variable types are atomic, and which are recursive.) The functions `is.recursive` and `is.atomic` let us test variables to see what type they are:

```
is.atomic(list())
## [1] FALSE
is.recursive(list())
## [1] TRUE
is.atomic(numeric())
## [1] TRUE
is.recursive(numeric())
## [1] FALSE
```

List Dimensions and Arithmetic

Like vectors, lists have a length. A list's length is the number of top-level elements that it contains:

```
length(a_list)
## [1] 4
length(main_list) #doesn't include the lengths of nested lists
## [1] 2
```

Again, like vectors, but unlike matrices, lists don't have dimensions. The `dim` function correspondingly returns `NULL`:

```
dim(a_list)
## NULL
```

`nrow`, `NROW`, and the corresponding column functions work on lists in the same way as on vectors:

```
nrow(a_list)
## NULL
ncol(a_list)
## NULL
NROW(a_list)
## [1] 4
```

```
NCOL(a_list)
```

```
## [1] 1
```

Unlike with vectors, arithmetic doesn't work on lists. Since each element can be of a different type, it doesn't make sense to be able to add or multiply two lists together. It is possible to do arithmetic on list elements, however, assuming that they are of an appropriate type. In that case, the usual rules for the element contents apply. For example:

```
l1 <- list(1:5)
l2 <- list(6:10)
l1[[1]] + l2[[1]]
```

```
## [1]  7  9 11 13 15
```

More commonly, you might want to perform arithmetic (or some other operation) on every element of a list. This requires looping, and will be discussed in Chapter 8.

Indexing Lists

Consider this test list:

```
l <- list(
  first  = 1,
  second = 2,
  third  = list(
    alpha = 3.1,
    beta  = 3.2
  )
)
```

As with vectors, we can access elements of the list using square brackets, [ ], and positive or negative numeric indices, element names, or a logical index. The following four lines of code all give the same result:

```
l[1:2]
```

```
## $first
## [1] 1
##
## $second
## [1] 2
```

```
l[-3]
```

```
## $first
## [1] 1
##
## $second
## [1] 2
```

```
l[c("first", "second")]
```

```
## $first
## [1] 1
##
## $second
## [1] 2
```

```
l[c(TRUE, TRUE, FALSE)]
```

```
## $first
## [1] 1
##
## $second
## [1] 2
```

The result of these indexing operations is *another list*. Sometimes we want to access the *contents* of the list elements instead. There are two operators to help us do this. Double square brackets ([[ ]]) can be given a single positive integer denoting the index to return, or a single string naming that element:

```
l[[1]]
```

```
## [1] 1
```

```
l[["first"]]
```

```
## [1] 1
```

The is.list function returns TRUE if the input is a list, and FALSE otherwise. For comparison, take a look at the two indexing operators:

```
is.list(l[1])
```

```
## [1] TRUE
```

```
is.list(l[[1]])
```

```
## [1] FALSE
```

For named elements of lists, we can also use the dollar sign operator, $. This works almost the same way as passing a named string to the double square brackets, but has two advantages. Firstly, many IDEs will autocomplete the name for you. (In R GUI, press Tab for this feature.) Secondly, R accepts partial matches of element names:

```
l$first
```

```
## [1] 1
```

```
l$f     #partial matching interprets "f" as "first"
```

```
## [1] 1
```

To access nested elements, we can stack up the square brackets or pass in a vector, though the latter method is less common and usually harder to read:

```
l[["third"]]["beta"]
```

```
## $beta
## [1] 3.2
```

```
l[["third"]][["beta"]]
```

```
## [1] 3.2
```

```
l[[c("third", "beta")]]
```

```
## [1] 3.2
```

The behavior when you try to access a nonexistent element of a list varies depending upon the type of indexing that you have used. For the next example, recall that our list, l, has only three elements.

If we use single square-bracket indexing, then the resulting list has an element with the value NULL (and name NA, if the original list has names). Compare this to bad indexing of a vector where the return value is NA:

```
l[c(4, 2, 5)]
```

```
## $<NA>
## NULL
##
## $second
## [1] 2
##
## $<NA>
## NULL
```

```
l[c("fourth", "second", "fifth")]
```

```
## $<NA>
## NULL
##
## $second
## [1] 2
##
## $<NA>
## NULL
```

Trying to access the contents of an element with an incorrect name, either with double square brackets or a dollar sign, returns NULL:

```
l[["fourth"]]
```

```
## NULL
```

```
l$fourth
```

```
## NULL
```

Finally, trying to access the contents of an element with an incorrect numerical index throws an error, stating that the subscript is out of bounds. This inconsistency in behavior is something that you just need to accept, though the best defense is to make sure that you check your indices before you use them:

```
l[[4]]        #this throws an error
```

Converting Between Vectors and Lists

Vectors can be converted to lists using the function `as.list`. This creates a list with each element of the vector mapping to a list element containing one value:

```
busy_beaver <- c(1, 6, 21, 107)  #See http://oeis.org/A060843
as.list(busy_beaver)

## [[1]]
## [1] 1
##
## [[2]]
## [1] 6
##
## [[3]]
## [1] 21
##
## [[4]]
## [1] 107
```

If each element of the list contains a scalar value, then it is also possible to convert that list to a vector using the functions that we have already seen (`as.numeric`, `as.charac ter`, and so on):

```
as.numeric(list(1, 6, 21, 107))

## [1]    1    6   21  107
```

This technique won't work in cases where the list contains nonscalar elements. This is a real issue, because as well as storing different types of data, lists are very useful for storing data of the same type, but with a nonrectangular shape:

```
(prime_factors <- list(
   two   = 2,
   three = 3,
   four  = c(2, 2),
   five  = 5,
   six   = c(2, 3),
   seven = 7,
   eight = c(2, 2, 2),
   nine  = c(3, 3),
   ten   = c(2, 5)
))
```

```
## $two
## [1] 2
##
## $three
## [1] 3
##
## $four
## [1] 2 2
##
## $five
## [1] 5
##
## $six
## [1] 2 3
##
## $seven
## [1] 7
##
## $eight
## [1] 2 2 2
##
## $nine
## [1] 3 3
##
## $ten
## [1] 2 5
```

This sort of list can be converted to a vector using the function unlist (it is sometimes technically possible to do this with mixed-type lists, but rarely useful):

```
unlist(prime_factors)

##    two  three  four1  four2   five   six1   six2  seven eight1 eight2
##      2      3      2      2      5      2      3      7      2      2
## eight3  nine1  nine2   ten1   ten2
##      2      3      3      2      5
```

Combining Lists

The c function that we have used for concatenating vectors also works for concatenating lists:

```
c(list(a = 1, b = 2), list(3))

## $a
## [1] 1
##
## $b
## [1] 2
##
## [[3]]
## [1] 3
```

If we use it to concatenate lists and vectors, the vectors are converted to lists (as though `as.list` had been called on them) before the concatenation occurs:

```
c(list(a = 1, b = 2), 3)
```

```
## $a
## [1] 1
##
## $b
## [1] 2
##
## [[3]]
## [1] 3
```

It is also possible to use the `cbind` and `rbind` functions on lists, but the resulting objects are very strange indeed. They are matrices with possibly nonscalar elements, or lists with dimensions, depending upon which way you want to look at them:

```
(matrix_list_hybrid <- cbind(
  list(a = 1, b = 2),
  list(c = 3, list(d = 4))
))
```

```
##   [,1] [,2]
## a 1    3
## b 2    List,1
```

```
str(matrix_list_hybrid)
```

```
## List of 4
##  $ : num 1
##  $ : num 2
##  $ : num 3
##  $ :List of 1
##   ..$ d: num 4
##  - attr(*, "dim")= int [1:2] 2 2
##  - attr(*, "dimnames")=List of 2
##   ..$ : chr [1:2] "a" "b"
##   ..$ : NULL
```

Using `cbind` and `rbind` in this way is something you shouldn't do often, and probably not at all. It's another case of R being a little too flexible and accommodating, instead of telling you that you've done something silly by throwing an error.

NULL

NULL is a special value that represents an empty variable. Its most common use is in lists, but it also crops up with data frames and function arguments. These other uses will be discussed later.

When you create a list, you may wish to specify that an element should exist, but should have no contents. For example, the following list contains UK bank holidays[1] for 2013 by month. Some months have no bank holidays, so we use NULL to represent this absence:

```
(uk_bank_holidays_2013 <- list(
  Jan = "New Year's Day",
  Feb = NULL,
  Mar = "Good Friday",
  Apr = "Easter Monday",
  May = c("Early May Bank Holiday", "Spring Bank Holiday"),
  Jun = NULL,
  Jul = NULL,
  Aug = "Summer Bank Holiday",
  Sep = NULL,
  Oct = NULL,
  Nov = NULL,
  Dec = c("Christmas Day", "Boxing Day")
))
## $Jan
## [1] "New Year's Day"
##
## $Feb
## NULL
##
## $Mar
## [1] "Good Friday"
##
## $Apr
## [1] "Easter Monday"
##
## $May
## [1] "Early May Bank Holiday" "Spring Bank Holiday"
##
## $Jun
## NULL
##
## $Jul
## NULL
##
## $Aug
## [1] "Summer Bank Holiday"
##
## $Sep
## NULL
##
## $Oct
## NULL
##
```

1. Bank holidays are public holidays.

```
## $Nov
## NULL
##
## $Dec
## [1] "Christmas Day" "Boxing Day"
```

It is important to understand the difference between NULL and the special missing value NA. The biggest difference is that NA is a scalar value, whereas NULL takes up no space at all—it has length zero:

```
length(NULL)
```

```
## [1] 0
```

```
length(NA)
```

```
## [1] 1
```

You can test for NULL using the function is.null. Missing values are not null:

```
is.null(NULL)
```

```
## [1] TRUE
```

```
is.null(NA)
```

```
## [1] FALSE
```

The converse test doesn't really make much sense. Since NULL has length zero, we have nothing to test to see if it is missing:

```
is.na(NULL)
```

```
## Warning: is.na() applied to non-(list or vector) of type 'NULL'
```

```
## logical(0)
```

NULL can also be used to remove elements of a list. Setting an element to NULL (even if it already contains NULL) will remove it. Suppose that for some reason we want to switch to an old-style Roman 10-month calendar, removing January and February:

```
uk_bank_holidays_2013$Jan <- NULL
uk_bank_holidays_2013$Feb <- NULL
uk_bank_holidays_2013
```

```
## $Mar
## [1] "Good Friday"
##
## $Apr
## [1] "Easter Monday"
##
## $May
## [1] "Early May Bank Holiday" "Spring Bank Holiday"
##
## $Jun
## NULL
##
```

```
## $Jul
## NULL
##
## $Aug
## [1] "Summer Bank Holiday"
##
## $Sep
## NULL
##
## $Oct
## NULL
##
## $Nov
## NULL
##
## $Dec
## [1] "Christmas Day" "Boxing Day"
```

To set an existing element to be NULL, we cannot simply assign the value of NULL, since that will remove the element. Instead, it must be set to list(NULL). Now suppose that the UK government becomes mean and cancels the summer bank holiday:

```
uk_bank_holidays_2013["Aug"] <- list(NULL)
uk_bank_holidays_2013

## $Mar
## [1] "Good Friday"
##
## $Apr
## [1] "Easter Monday"
##
## $May
## [1] "Early May Bank Holiday" "Spring Bank Holiday"
##
## $Jun
## NULL
##
## $Jul
## NULL
##
## $Aug
## NULL
##
## $Sep
## NULL
##
## $Oct
## NULL
##
## $Nov
## NULL
##
```

```
## $Dec
## [1] "Christmas Day" "Boxing Day"
```

Pairlists

R has another sort of list, the *pairlist*. Pairlists are used internally to pass arguments into functions, but you should almost never have to actively use them. Possibly the only time[2] that you are likely to explicitly see a pairlist is when using `formals`. That function returns a pairlist of the arguments of a function.

Looking at the help page for the standard deviation function, `?sd`, we see that it takes two arguments, a vector x and a logical value `na.rm`, which has a default value of `FALSE`:

```
(arguments_of_sd <- formals(sd))

## $x
##
##
## $na.rm
## [1] FALSE

class(arguments_of_sd)

## [1] "pairlist"
```

For most practical purposes, pairlists behave like lists. The only difference is that a pairlist of length zero is `NULL`, but a list of length zero is just an empty list:

```
pairlist()

## NULL

list()

## list()
```

Data Frames

Data frames are used to store spreadsheet-like data. They can either be thought of as matrices where each column can store a different type of data, or nonnested lists where each element is of the same length.

---

2. R also stores some global settings in a pairlist variable called `.Options` in the base environment. You shouldn't access this variable directly, but instead use the function `options`, which returns a list.

Creating Data Frames

We create data frames with the `data.frame` function:

```
(a_data_frame <- data.frame(
  x = letters[1:5],
  y = rnorm(5),
  z = runif(5) > 0.5
))
```

```
##   x        y    z
## 1 a  0.17581 TRUE
## 2 b  0.06894 TRUE
## 3 c  0.74217 TRUE
## 4 d  0.72816 TRUE
## 5 e -0.28940 TRUE
```

```
class(a_data_frame)
```

```
## [1] "data.frame"
```

Notice that each column can have a different type than the other columns, but that all the elements within a column are the same type. Also notice that the class of the object is `data.frame`, with a dot rather than a space.

In this example, the rows have been automatically numbered from one to five. If any of the input vectors had names, then the row names would have been taken from the first such vector. For example, if y had names, then those would be given to the data frame:

```
y <- rnorm(5)
names(y) <- month.name[1:5]
data.frame(
  x = letters[1:5],
  y = y,
  z = runif(5) > 0.5
)
```

```
##             x       y     z
## January     a -0.9373 FALSE
## February    b  0.7314  TRUE
## March       c -0.3030  TRUE
## April       d -1.3307 FALSE
## May         e -0.6857 FALSE
```

This behavior can be overridden by passing the argument `row.names = NULL` to the `data.frame` function:

```
data.frame(
  x = letters[1:5],
  y = y,
  z = runif(5) > 0.5,
  row.names = NULL
)
```

```
##   x       y     z
## 1 a -0.9373 FALSE
## 2 b  0.7314 FALSE
## 3 c -0.3030  TRUE
## 4 d -1.3307  TRUE
## 5 e -0.6857 FALSE
```

It is also possible to provide your own row names by passing a vector to row.names. This vector will be converted to character, if it isn't already that type:

```
data.frame(
  x = letters[1:5],
  y = y,
  z = runif(5) > 0.5,
  row.names = c("Jackie", "Tito", "Jermaine", "Marlon", "Michael")
)
```

```
##           x       y     z
## Jackie    a -0.9373  TRUE
## Tito      b  0.7314 FALSE
## Jermaine  c -0.3030  TRUE
## Marlon    d -1.3307 FALSE
## Michael   e -0.6857 FALSE
```

The row names can be retrieved or changed at a later date, in the same manner as with matrices, using rownames (or row.names). Likewise, colnames and dimnames can be used to get or set the column and dimension names, respectively. In fact, more or less all the functions that can be used to inspect matrices can also be used with data frames. nrow, ncol, and dim also work in exactly the same way as they do in matrices:

```
rownames(a_data_frame)
```

```
## [1] "1" "2" "3" "4" "5"
```

```
colnames(a_data_frame)
```

```
## [1] "x" "y" "z"
```

```
dimnames(a_data_frame)
```

```
## [[1]]
## [1] "1" "2" "3" "4" "5"
##
## [[2]]
## [1] "x" "y" "z"
```

```
nrow(a_data_frame)
```

```
## [1] 5
```

```
ncol(a_data_frame)
```

```
## [1] 3
```

```
dim(a_data_frame)
```

```
## [1] 5 3
```

There are two quirks that you need to be aware of. First, length returns the same value as ncol, *not* the total number of elements in the data frame. Likewise, names returns the same value as colnames. For clarity of code, I recommend that you avoid these two functions, and use ncol and colnames instead:

```
length(a_data_frame)
## [1] 3
names(a_data_frame)
## [1] "x" "y" "z"
```

It is possible to create a data frame by passing different lengths of vectors, as long as the lengths allow the shorter ones to be recycled an exact number of times. More technically, the lowest common multiple of all the lengths must be equal to the longest vector:

```
data.frame(       #lengths 1, 2, and 4 are OK
  x = 1,          #recycled 4 times
  y = 2:3,        #recycled twice
  z = 4:7         #the longest input; no recycling
)
##   x y z
## 1 1 2 4
## 2 1 3 5
## 3 1 2 6
## 4 1 3 7
```

If the lengths are not compatible, then an error will be thrown:

```
data.frame(       #lengths 1, 2, and 3 cause an error
  x = 1,          #lowest common multiple is 6, which is more than 3
  y = 2:3,
  z = 4:6
)
```

One other consideration when creating data frames is that by default the column names are checked to be unique, valid variable names. This feature can be turned off by passing check.names = FALSE to data.frame:

```
data.frame(
  "A column"    = letters[1:5],
  "!@#$%^&*()"  = rnorm(5),
  "..."         = runif(5) > 0.5,
  check.names   = FALSE
)
##   A column !@#$%^&*()   ...
## 1        a    0.32940  TRUE
## 2        b   -1.81969  TRUE
## 3        c    0.22951 FALSE
## 4        d   -0.06705  TRUE
## 5        e   -1.58005  TRUE
```

In general, having nonstandard column names is a bad idea. Duplicating column names is even worse, since it can lead to hard-to-find bugs once you start taking subsets. Turn off the column name checking at your own peril.

Indexing Data Frames

There are lots of different ways of indexing a data frame. To start with, pairs of the four different vector indices (positive integers, negative integers, logical values, and characters) can be used in exactly the same way as with matrices. These commands both select the second and third elements of the first two columns:

```
a_data_frame[2:3, -3]

##   x       y
## 2 b 0.06894
## 3 c 0.74217

a_data_frame[c(FALSE, TRUE, TRUE, FALSE, FALSE), c("x", "y")]

##   x       y
## 2 b 0.06894
## 3 c 0.74217
```

Since more than one column was selected, the resultant subset is also a data frame. If only one column had been selected, the result would have been simplified to be a vector:

```
class(a_data_frame[2:3, -3])

## [1] "data.frame"

class(a_data_frame[2:3, 1])

## [1] "factor"
```

If we only want to select one column, then list-style indexing (double square brackets with a positive integer or name, or the dollar sign operator with a name) can also be used. These commands all select the second and third elements of the first column:

```
a_data_frame$x[2:3]

## [1] b c
## Levels: a b c d e

a_data_frame[[1]][2:3]

## [1] b c
## Levels: a b c d e

a_data_frame[["x"]][2:3]

## [1] b c
## Levels: a b c d e
```

If we are trying to subset a data frame by placing conditions on columns, the syntax can get a bit clunky, and the subset function provides a cleaner alternative. subset takes up to three arguments: a data frame to subset, a logical vector of conditions for rows to include, and a vector of column names to keep (if this last argument is omitted, then all the columns are kept). The genius of subset is that it uses special evaluation techniques to let you avoid doing some typing: instead of you having to type a_data_frame$y to access the y column of a_data_frame, it already knows which data frame to look in, so you can just type y. Likewise, when selecting columns, you don't need to enclose the names of the columns in quotes; you can just type the names directly. In this next example, recall that | is the operator for logical *or*:

```
a_data_frame[a_data_frame$y > 0 | a_data_frame$z, "x"]

## [1] a b c d e
## Levels: a b c d e

subset(a_data_frame, y > 0 | z, x)

##   x
## 1 a
## 2 b
## 3 c
## 4 d
## 5 e
```

Basic Data Frame Manipulation

Like matrices, data frames can be transposed using the t function, but in the process all the columns (which become rows) are converted to the same type, and the whole thing becomes a matrix:

```
t(a_data_frame)

##   [,1]        [,2]       [,3]       [,4]       [,5]
## x "a"         "b"        "c"        "d"        "e"
## y " 0.17581"  " 0.06894" " 0.74217" " 0.72816" "-0.28940"
## z "TRUE"      "TRUE"     "TRUE"     "TRUE"     "TRUE"
```

Data frames can also be joined together using cbind and rbind, assuming that they have the appropriate sizes. rbind is smart enough to reorder the columns to match. cbind doesn't check column names for duplicates, though, so be careful with it:

```
another_data_frame <- data.frame(  #same cols as a_data_frame, different order
  z = rlnorm(5),                   #lognormally distributed numbers
  y = sample(5),                   #the numbers 1 to 5, in some order
  x = letters[3:7]
)
rbind(a_data_frame, another_data_frame)
```

```
##    x        y       z
## 1  a  0.17581 1.0000
## 2  b  0.06894 1.0000
## 3  c  0.74217 1.0000
## 4  d  0.72816 1.0000
## 5  e -0.28940 1.0000
## 6  c  1.00000 0.8714
## 7  d  3.00000 0.2432
## 8  e  5.00000 2.3498
## 9  f  4.00000 2.0263
## 10 g  2.00000 1.7145

cbind(a_data_frame, another_data_frame)

##    x        y    z      z y x
## 1  a  0.17581 TRUE 0.8714 1 c
## 2  b  0.06894 TRUE 0.2432 3 d
## 3  c  0.74217 TRUE 2.3498 5 e
## 4  d  0.72816 TRUE 2.0263 4 f
## 5  e -0.28940 TRUE 1.7145 2 g
```

Where two data frames share columns, they can be merged together using the merge function. merge provides a variety of options for doing database-style joins. To join two data frames, you need to specify which columns contain the key values to match up. By default, the merge function uses all the common columns from the two data frames, but more commonly you will just want to use a single shared ID column. In the following examples, we specify that the x column contains our IDs using the by argument:

```
merge(a_data_frame, another_data_frame, by = "x")

##    x     y.x  z.x    z.y y.y
## 1  c  0.7422 TRUE 0.8714   1
## 2  d  0.7282 TRUE 0.2432   3
## 3  e -0.2894 TRUE 2.3498   5

merge(a_data_frame, another_data_frame, by = "x", all = TRUE)

##    x      y.x  z.x    z.y y.y
## 1  a  0.17581 TRUE     NA  NA
## 2  b  0.06894 TRUE     NA  NA
## 3  c  0.74217 TRUE 0.8714   1
## 4  d  0.72816 TRUE 0.2432   3
## 5  e -0.28940 TRUE 2.3498   5
## 6  f      NA   NA 2.0263   4
## 7  g      NA   NA 1.7145   2
```

Where a data frame has all numeric values, the functions colSums and colMeans can be used to calculate the sums and means of each column, respectively. Similarly, rowSums and rowMeans calculate the sums and means of each row:

```
colSums(a_data_frame[, 2:3])

##     y     z
## 1.426 5.000
```

```
colMeans(a_data_frame[, 2:3])
##      y      z
## 0.2851 1.0000
```

Manipulating data frames is a huge topic, and is covered in more depth in Chapter 13.

Summary

- Lists can contain different sizes and types of variables in each element.
- Lists are recursive variables, since they can contain other lists.
- You can index lists using [ ], [[ ]], or $.
- NULL is a special value that can be used to create "empty" list elements.
- Data frames store spreadsheet-like data.
- Data frames have some properties of matrices (they are rectangular), and some of lists (different columns can contain different sorts of variables).
- Data frames can be indexed like matrices or like lists.
- merge lets you do database-style joins on data frames.

Test Your Knowledge: Quiz

Question 5-1

What is the length of this list?

```
list(alpha = 1, list(beta = 2, gamma = 3, delta = 4), eta = NULL)
## $alpha
## [1] 1
##
## [[2]]
## [[2]]$beta
## [1] 2
##
## [[2]]$gamma
## [1] 3
##
## [[2]]$delta
## [1] 4
##
##
## $eta
## NULL
```

Question 5-2

Where might you find a pairlist being used?

Question 5-3

Name as many ways as you can think of to create a subset of a data frame.

Question 5-4

How would you create a data frame where the column names weren't unique, valid variable names?

Question 5-5

Which function would you use to append one data frame to another?

Test Your Knowledge: Exercises

Exercise 5-1

Create a list variable that contains all the square numbers in the range 0 to 9 in the first element, in the range 10 to 19 in the second element, and so on, up to a final element with square numbers in the range 90 to 99. Elements with no square numbers should be included! [10]

Exercise 5-2

R ships with several built-in datasets, including the famous[3] iris (flowers, not eyes) data collected by Anderson and analyzed by Fisher in the 1930s. Type iris to see the dataset. Create a new data frame that consists of the numeric columns of the iris dataset, and calculate the means of its columns. [5]

Exercise 5-3

The beaver1 and beaver2 datasets contain body temperatures of two beavers. Add a column named id to the beaver1 dataset, where the value is always 1. Similarly, add an id column to beaver2, with value 2. Vertically concatenate the two data frames and find the subset where either beaver is active. [10]

---

3. By some definitions of fame.

## CHAPTER 6

Environments and Functions

We've already used a variety of the functions that come with R. In this chapter, you'll learn what a function is, and how to write your own. Before that, we'll take a look at environments, which are used to store variables.

Chapter Goals

After reading this chapter, you should:

- Know what an environment is, and how to create one
- Be able to create, access, and list variables within an environment
- Understand the components that make up a function
- Be able to write your own functions
- Understand variable scope

Environments

All the variables that we create need to be stored somewhere, and that somewhere is an environment. Environments themselves are just another type of variable—we can assign them, manipulate them, and pass them into functions as arguments, just like we would any other variable. They are closely related to lists in that they are used for storing different types of variables together. In fact, most of the syntax for lists also works for environments, and we can coerce a list to be an environment (and vice versa).

Usually, you won't need to explicitly deal with environments. For example, when you assign a variable at the command prompt, it will automatically go into an environment called the *global environment* (also known as the *user workspace*). When you call a function, an environment is automatically created to store the function-related variables. Understanding the basics of environments can be useful, however, in understanding the scope of variables, and for examining the call stack when debugging your code.

Slightly annoyingly, environments aren't created with the `environment` function (that function returns the environment that contains a particular function). Instead, what we want is `new.env`:

```
an_environment <- new.env()
```

Assigning variables into environments works in exactly the same way as with lists. You can either use double square brackets or the dollar sign operator. As with lists, the variables can be of different types and sizes:

```
an_environment[["pythag"]] <- c(12, 15, 20, 21) #See http://oeis.org/A156683
an_environment$root <- polyroot(c(6, -5, 1))
```

The `assign` function that we saw in "Assigning Variables" on page 17 takes an optional environment argument that can be used to specify where the variable is stored:

```
assign(
  "moonday",
  weekdays(as.Date("1969/07/20")),
  an_environment
)
```

Retrieving the variables works in the same way—you can either use list-indexing syntax, or `assign`'s opposite, the `get` function:

```
an_environment[["pythag"]]
```

```
## [1] 12 15 20 21
```

```
an_environment$root
```

```
## [1] 2+0i 3-0i
```

```
get("moonday", an_environment)
```

```
## [1] "Sunday"
```

The `ls` and `ls.str` functions also take an environment argument, allowing you to list their contents:

```
ls(envir = an_environment)
```

```
## [1] "moonday" "pythag"  "root"
```

```
ls.str(envir = an_environment)
```

```
## moonday :  chr "Sunday"
## pythag :  num [1:4] 12 15 20 21
## root :  cplx [1:2] 2+0i 3-0i
```

We can test to see if a variable exists in an environment using the `exists` function:

```
exists("pythag", an_environment)
```

```
## [1] TRUE
```

Conversion from environment to list and back again uses the obvious functions, as.list and as.environment. In the latter case, there is also a function list2env that allows for a little more flexibility in the creation of the environment:

```
#Convert to list
(a_list <- as.list(an_environment))
```

```
## $pythag
## [1] 12 15 20 21
##
## $moonday
## [1] "Sunday"
##
## $root
## [1] 2+0i 3-0i
```

```
#...and back again.  Both lines of code do the same thing.
as.environment(a_list)
```

```
## <environment: 0x000000004a6fe290>
```

```
list2env(a_list)
```

```
## <environment: 0x000000004ad10288>
```

All environments are nested, meaning that they must have a parent environment (the exception is a special environment called the *empty environment* that sits at the top of the chain). By default, the exists and get functions will also look for variables in the parent environments. Pass inherits = FALSE to them to change this behavior so that they will only look in the environment that you've specified:

```
nested_environment <- new.env(parent = an_environment)
exists("pythag", nested_environment)
```

```
## [1] TRUE
```

```
exists("pythag", nested_environment, inherits = FALSE)
```

```
## [1] FALSE
```

 The word "frame" is used almost interchangeably with "environment." (See section 2.1.10 of the R Language Definition manual that ships with R for the technicalities.) This means that some functions that work with environments have "frame" in their name, parent.frame being the most common of these.

Shortcut functions are available to access both the global environment (where variables that you assign from the command prompt are stored) and the base environment (this contains functions and other variables from R's base package, which provides basic functionality):

```
non_stormers <<- c(3, 7, 8, 13, 17, 18, 21) #See http://oeis.org/A002312
get("non_stormers", envir = globalenv())

## [1]  3  7  8 13 17 18 21

head(ls(envir = baseenv()), 20)

##  [1] "-"                 "-.Date"             "-.POSIXt"
##  [4] "!"                 "!.hexmode"          "!.octmode"
##  [7] "!="               "$"                  "$.data.frame"
## [10] "$.DLLInfo"         "$.package_version"  "$<-"
## [13] "$<-.data.frame"    "%%"                 "%*%"
## [16] "%/%"               "%in%"               "%o%"
## [19] "%x%"               "&"
```

There are two other situations where we might encounter environments. First, whenever a function is called, all the variables defined by the function are stored in an environment belonging to that function (a function plus its environment is sometimes called a *closure*). Second, whenever we load a package, the functions in that package are stored in an environment on the search path. This will be discussed in Chapter 10.

Functions

While most variable types are for storing data, functions let us *do* things with data—they are "verbs" rather than "nouns." Like environments, they are just another data type that we can assign and manipulate and even pass into other functions.

Creating and Calling Functions

In order to understand functions better, let's take a look at what they consist of.

Typing the name of a function shows you the code that runs when you call it. This is the rt function, which generates random numbers from a t-distribution:[1]

```
rt

## function (n, df, ncp)
## {
##     if (missing(ncp))
##         .External(C_rt, n, df)
##     else rnorm(n, ncp)/sqrt(rchisq(n, df)/df)
## }
```

---

1. If the definition is a single line that says something like UseMethod("my_function") or standardGene ric("my_function"), see "Object-Oriented Programming" on page 302 in Chapter 16. If R complains that the object is not found, try getAnywhere(my_function).

```
## <bytecode: 0x0000000019738e10>
## <environment: namespace:stats>
```

As you can see, rt takes up to three input arguments: n is the number of random numbers to generate, df is the number of degrees of freedom, and ncp is an optional noncentrality parameter. To be technical, the three arguments n, df, and ncp are the *formal arguments* of rt. When you are calling the function and passing values to it, those values are just called *arguments*.

 The difference between arguments and formal arguments isn't very important, so the rest of the book doesn't make an effort to differentiate between the two concepts.

In between the curly braces, you can see the lines of code that constitute the *body* of the function. This is the code that is executed each time you call rt.

Notice that there is no explicit "return" keyword to state which value should be returned from the function. In R, the last value that is calculated in the function is automatically returned. In the case of rt, if the ncp argument is omitted, some C code is called to generate the random numbers, and those are returned. Otherwise, the function calls the rnorm, rchisq, and sqrt functions to generate the numbers, and those are returned.

To create our own functions, we just assign them as we would any other variable. As an example, let's create a function to calculate the length of the hypotenuse of a right-angled triangle (for simplicity, we'll use the obvious algorithm; for real-world code, this doesn't work well with very big and very small numbers, so you shouldn't calculate hypotenuses this way):

```
hypotenuse <- function(x, y)
{
  sqrt(x ^ 2 + y ^ 2)
}
```

Here, hypotenuse is the name of the function we are creating, x and y are its (*formal*) arguments, and the contents of the braces are the function body.

Actually, since our function body is only one line of code, we can omit the braces:

```
hypotenuse <- function(x, y) sqrt(x ^ 2 + y ^ 2)   #same as before
```

R is very permissive about how you space your code, so "one line of code" can be stretched to run over several lines. The amount of code that can be included without braces is one statement. The exact definition of a statement is technical, but from a practical point of view, it is the amount of code that you can type at the command line before it executes.

We can now call this function as we would any other:

```
hypotenuse(3, 4)
## [1] 5
hypotenuse(y = 24, x = 7)
## [1] 25
```

When we call a function, if we don't name the arguments, then R will match them based on position. In the case of hypotenuse(3, 4), 3 comes first so it is mapped to x, and 4 comes second so it is mapped to y.

If we want to change the order that we pass the arguments, or omit some of them, then we can pass named arguments. In the case of hypotenuse(y = 24, x = 7), although we pass the variables in the "wrong" order, R still correctly determines which variable should be mapped to x, and which to y.

It doesn't make much sense for a hypotenuse-calculating function, but if we wanted, we could provide default values for x and y. In this new version, if we don't pass anything to the function, x takes the value 5 and y takes the value 12:

```
hypotenuse <- function(x = 5, y = 12)
{
  sqrt(x ^ 2 + y ^ 2)
}
hypotenuse() #equivalent to hypotenuse(5, 12)
## [1] 13
```

We've already seen the formals function for retrieving the arguments of a function as a (pair)list. The args function does the same thing in a more human-readable, but less programming-friendly, way. formalArgs returns a character vector of the names of the arguments:

```
formals(hypotenuse)
## $x
## [1] 5
##
## $y
## [1] 12
```

```
args(hypotenuse)
```

```
## function (x = 5, y = 12)
## NULL
```

```
formalArgs(hypotenuse)
```

```
## [1] "x" "y"
```

The body of a function is retrieved using the body function. This isn't often very useful on its own, but we may sometimes want to examine it as text—to find functions that call another function, for example. We can use deparse to achieve this:

```
(body_of_hypotenuse <- body(hypotenuse))
```

```
## {
##     sqrt(x^2 + y^2)
## }
```

```
deparse(body_of_hypotenuse)
```

```
## [1] "{"                 "    sqrt(x^2 + y^2)" "}"
```

The default values given to formal arguments of functions can be more than just constant values—we can pass any R code into them, and even use other formal arguments. The following function, normalize, scales a vector. The arguments m and s are, by default, the mean and standard deviation of the first argument, so that the returned vector will have mean 0 and standard deviation 1:

```
normalize <- function(x, m = mean(x), s = sd(x))
{
  (x - m) / s
}
normalized <- normalize(c(1, 3, 6, 10, 15))
mean(normalized)        #almost 0!
```

```
## [1] -5.573e-18
```

```
sd(normalized)
```

```
## [1] 1
```

There is a little problem with our normalize function, though, which we can see if some of the elements of x are missing:

```
normalize(c(1, 3, 6, 10, NA))
```

```
## [1] NA NA NA NA NA
```

If any elements of a vector are missing, then by default, mean and sd will both return NA. Consequently, our normalize function returns NA values everywhere. It might be preferable to have the option of only returning NA values where the input was NA. Both mean and sd have an argument, na.rm, that lets us remove missing values before any calculations occur. To avoid all the NA values, we could include such an argument in normalize:

```
normalize <- function(x, m = mean(x, na.rm = na.rm),
  s = sd(x, na.rm = na.rm), na.rm = FALSE)
{
  (x - m) / s
}
normalize(c(1, 3, 6, 10, NA))
```

```
## [1] NA NA NA NA NA
```

```
normalize(c(1, 3, 6, 10, NA), na.rm = TRUE)
```

```
## [1] -1.0215 -0.5108  0.2554  1.2769      NA
```

This works, but the syntax is a little clunky. To save us having to explicitly type the names of arguments that aren't actually used by the function (na.rm is only being passed to mean and sd), R has a special argument, ..., that contains all the arguments that aren't matched by position or name:

```
normalize <- function(x, m = mean(x, ...), s = sd(x, ...), ...)
{
  (x - m) / s
}
normalize(c(1, 3, 6, 10, NA))
```

```
## [1] NA NA NA NA NA
```

```
normalize(c(1, 3, 6, 10, NA), na.rm = TRUE)
```

```
## [1] -1.0215 -0.5108  0.2554  1.2769      NA
```

Now in the call normalize(c(1, 3, 6, 10, NA), na.rm = TRUE), the argument na.rm does not match any of the formal arguments of normalize, since it isn't x or m or s. That means that it gets stored in the ... argument of normalize. When we evaluate m, the expression mean(x, ...) is now mean(x, na.rm = TRUE).

If this isn't clear right now, don't worry. How this works is an advanced topic, and most of the time we don't need to worry about it. For now, you just need to know that ... can be used to pass arguments to subfunctions.

Passing Functions to and from Other Functions

Functions can be used just like other variable types, so we can pass them as arguments to other functions, and return them from functions. One common example of a function that takes another function as an argument is do.call. This function provides an alternative syntax for calling other functions, letting us pass the arguments as a list, rather than one at a time:

```
do.call(hypotenuse, list(x = 3, y = 4)) #same as hypotenuse(3, 4)
```

```
## [1] 5
```

Perhaps the most common use case for do.call is with rbind. You can use these two functions together to concatenate several data frames or matrices together at once:

```
dfr1 <- data.frame(x = 1:5, y = rt(5, 1))
dfr2 <- data.frame(x = 6:10, y = rf(5, 1, 1))
dfr3 <- data.frame(x = 11:15, y = rbeta(5, 1, 1))
do.call(rbind, list(dfr1, dfr2, dfr3)) #same as rbind(dfr1, dfr2, dfr3)
```

```
##      x        y
## 1   1  1.10440
## 2   2  0.87931
## 3   3 -1.18288
## 4   4 -1.04847
## 5   5  0.90335
## 6   6  0.27186
## 7   7  2.49953
## 8   8  0.89534
## 9   9  4.21537
## 10 10  0.07751
## 11 11  0.31153
## 12 12  0.29114
## 13 13  0.01079
## 14 14  0.97188
## 15 15  0.53498
```

It is worth spending some time getting comfortable with this idea. In Chapter 9, we're going to make a lot of use of passing functions to other functions with apply and its derivatives.

When using functions as arguments, it isn't necessary to assign them first. In the same way that we could simplify this:

```
menage <- c(1, 0, 0, 1, 2, 13, 80) #See http://oeis.org/A000179
mean(menage)
```

```
## [1] 13.86
```

to:

```
mean(c(1, 0, 0, 1, 2, 13, 80))
```

```
## [1] 13.86
```

we can also pass functions *anonymously*:

```
x_plus_y <- function(x, y) x + y
do.call(x_plus_y, list(1:5, 5:1))
```

```
## [1] 6 6 6 6 6
```

```
#is the same as
do.call(function(x, y) x + y, list(1:5, 5:1))
```

```
## [1] 6 6 6 6 6
```

Functions that return functions are rarer, but no less valid for it. The ecdf function returns the empirical cumulative distribution function of a vector, as seen in Figure 6-1:

```
(emp_cum_dist_fn <- ecdf(rnorm(50)))
```

```
## Empirical CDF
## Call: ecdf(rnorm(50))
##   x[1:50] = -2.2, -2.1,  -2,  ..., 1.9, 2.6
```

```
is.function(emp_cum_dist_fn)
```

```
## [1] TRUE
```

```
plot(emp_cum_dist_fn)
```

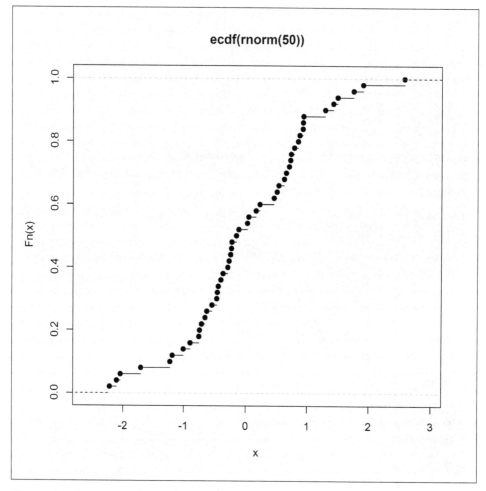

Figure 6-1. An empirical cumulative distribution function

Variable Scope

A variable's *scope* is the set of places from which you can see the variable. For example, when you define a variable inside a function, the rest of the statements in that function will have access to that variable. In R (but not S), subfunctions will also have access to that variable. In this next example, the function f takes a variable x and passes it to the function g. f also defines a variable y, which is within the scope of g, since g is a sub-function of f. So, even though y isn't defined inside g, the example works:

```
f <- function(x)
{
  y <- 1
  g <- function(x)
  {
    (x + y) / 2 #y is used, but is not a formal argument of g
  }
  g(x)
}
f(sqrt(5))      #It works! y is magically found in the environment of f

## [1] 1.618
```

If we modify the example to define g outside of f, so it is not a subfunction of f, the example will throw an error, since R cannot find y:

```
f <- function(x)
{
  y <- 1
  g(x)
}
g <- function(x)
{
  (x + y) / 2
}
f(sqrt(5))

## January February   March   April     May
## 0.6494   1.4838   0.9665  0.4527  0.7752
```

In the section "Environments" on page 79, we saw that the get and exists functions look for variables in parent environments as well as the current one. Variable scope works in exactly the same way: R will try to find variables in the current environment, and if it doesn't find them it will look in the parent environment, and then that environment's parent, and so on until it reaches the global environment. Variables defined in the global environment can be seen from anywhere, which is why they are called global variables.

In our first example, the environment belonging to f is the parent environment of the environment belonging to g, which is why y can be found. In the second example, the parent environment of g is the global environment, which doesn't contain a variable y, which is why an error is thrown.

This system of scoping where variables can be found in parent environments is often useful, but also brings the potential for mischief and awful, unmaintainable code. Consider the following function, h:

```
h <- function(x)
{
  x * y
}
```

It looks like it shouldn't work, since it accepts a single argument, x, but uses two arguments, x and y, in its body. Let's try it, with a clean user workspace:

```
h(9)
```

```
##   January February   March   April    May
##    -8.436    6.583  -2.727 -11.976  -6.171
```

So far, our intuition holds. y is not defined, so the function throws an error. Now look at what happens if we define y in the user workspace:

```
y <- 16
h(9)
```

```
## [1] 144
```

When R fails to find a variable named y in the environment belonging to h, it looks in h's parent—the user workspace (a.k.a. global environment), where y is defined—and the product is correctly calculated.

Global variables should be used sparingly, since they make it very easy to write appalling code. In this modified function, h2, y is randomly locally defined half the time. With y defined in the user workspace, when we evaluate it y will be randomly local or global!

```
h2 <- function(x)
{
  if(runif(1) > 0.5) y <- 12
  x * y
}
```

Let's use `replicate` to run the code several times to see the result:

```
replicate(10, h2(9))
```

```
## [1] 144 144 144 108 144 108 108 144 108 108
```

When the uniform random number (between 0 and 1) generated by `runif` is greater than 0.5, a local variable y is assigned the value 12. Otherwise, the global value of 16 is used.

As I'm sure you've noticed, it is very easy to create obscure bugs in code by doing things like this. Usually it is better to explicitly pass all the variables that we need into a function.

Summary

- Environments store variables and can be created with new.env.
- You can treat environments like lists most of the time.
- All environments have a parent environment (except the empty environment at the top).
- Functions consist of formal arguments and a body.
- You can assign and use functions just as you would any other variable type.
- R will look for variables in the current environment and its parents.

Test Your Knowledge: Quiz

Question 6-1
What is another name for the global environment?

Question 6-2
How would you convert a list to an environment?

Question 6-3
How do you print the contents of a function to the console?

Question 6-4
Name three functions that tell you the names of the formal arguments of a function.

Question 6-5
What does the do.call function do?

Test Your Knowledge: Exercises

Exercise 6-1
Create a new environment named multiples_of_pi. Assign these variables into the environment:

1. two_pi, with the value 2 * n, using double square brackets
2. three_pi, with the value 3 * n, using the dollar sign operator
3. four_pi, with the value 4 * n, using the assign function

   List the contents of the environment, along with their values. [10]

Exercise 6-2

Write a function that accepts a vector of integers (for simplicity, you don't *have* to worry about input checking) and returns a logical vector that is TRUE whenever the input is even, FALSE whenever the input is odd, and NA whenever the input is non-finite (nonfinite means anything that will make is.finite return FALSE: Inf, -Inf, NA, and NaN). Check that the function works with positive, negative, zero, and non-finite inputs. [10]

Exercise 6-3

Write a function that accepts a function as an input and returns a list with two elements: an element named args that contains a pairlist of the input's formal arguments, and an element named body that contains the input's body. Test it by calling the function with a variety of inputs. [10]

## CHAPTER 7

Strings and Factors

As well as dealing with numbers and logical values, at some point you will almost certainly need to manipulate text. This is particularly common when you are retrieving or cleaning datasets. Perhaps you are trying to turn the text of a log file into meaningful values, or correct the typos in your data. These data-cleaning activities will be discussed in more depth in Chapter 13, but for now, you will learn how to manipulate character vectors.

Factors are used to store categorical data like gender ("male" or "female") where there are a limited number of options for a string. They sometimes behave like character vectors and sometimes like integer vectors, depending upon context.

Chapter Goals

After reading this chapter, you should:

- Be able to construct new strings from existing strings
- Be able to format how numbers are printed
- Understand special characters like tab and newline
- Be able to create and manipulate factors

Strings

Text data is stored in character vectors (or, less commonly, character arrays). It's important to remember that each element of a character vector is a whole string, rather than just an individual character. In R, "string" is an informal term that is used because "element of a character vector" is quite a mouthful.

The fact that the basic unit of text is a character *vector* means that most string manipulation functions operate on vectors of strings, in the same way that mathematical operations are vectorized.

Constructing and Printing Strings

As you've already seen, character vectors can be created with the c function. We can use single or double quotes around our strings, as long as they match, though double quotes are considered more standard:

```
c(
  "You should use double quotes most of the time",
  'Single quotes are better for including " inside the string'
)
## [1] "You should use double quotes most of the time"
## [2] "Single quotes are better for including \" inside the string"
```

The paste function combines strings together. Each vector passed to it has its elements recycled to reach the length of the longest input, and then the strings are concatenated, with a space separating them. We can change the separator by passing an argument called sep, or use the related function paste0 to have no separator. After all the strings are combined, the result can be collapsed into one string containing everything using the collapse argument:

```
paste(c("red", "yellow"), "lorry")

## [1] "red lorry"    "yellow lorry"

paste(c("red", "yellow"), "lorry", sep = "-")

## [1] "red-lorry"    "yellow-lorry"

paste(c("red", "yellow"), "lorry", collapse = ", ")

## [1] "red lorry, yellow lorry"

paste0(c("red", "yellow"), "lorry")

## [1] "redlorry"    "yellowlorry"
```

The function toString is a variation of paste that is useful for printing vectors. It separates each element with a comma and a space, and can limit how much we print. In the following example, width = 40 limits the output to 40 characters:

```
x <- (1:15) ^ 2
toString(x)

## [1] "1, 4, 9, 16, 25, 36, 49, 64, 81, 100, 121, 144, 169, 196, 225"

toString(x, width = 40)

## [1] "1, 4, 9, 16, 25, 36, 49, 64, 81, 100...."
```

cat is a low-level function that works similarly to paste, but with less formatting. You should rarely need to call it directly yourself, but it is worth being aware of, since it is the basis for most of the print functions. cat can also accept a file argument[1] to write its output to a file:

```
cat(c("red", "yellow"), "lorry")
## red yellow lorry
```

Usually, when strings are printed to the console they are shown wrapped in double quotes. By wrapping a variable in a call to the noquote function, we can suppress those quotes. This can make the text more readable in some instances:

```
x <- c(
  "I", "saw", "a", "saw", "that", "could", "out",
  "saw", "any", "other", "saw", "I", "ever", "saw"
)
y <- noquote(x)
x
## [1] "I"     "saw"    "a"     "saw"   "that"  "could" "out"   "saw"
## [9] "any"   "other" "saw"   "I"     "ever"  "saw"

y
## [1] I     saw   a     saw   that could out   saw   any   other saw
## [12] I    ever  saw
```

Formatting Numbers

There are several functions for formatting numbers. formatC uses C-style formatting specifications that allow you to specify fixed or scientific formatting, the number of decimal places, and the width of the output. Whatever the options, the input should be one of the numeric types (including arrays), and the output is a character vector or array:

```
pow <- 1:3
(powers_of_e <- exp(pow))
## [1]  2.718  7.389 20.086

formatC(powers_of_e)
## [1] "2.718" "7.389" "20.09"

formatC(powers_of_e, digits = 3)          #3 sig figs
## [1] "2.72" "7.39" "20.1"

formatC(powers_of_e, digits = 3, width = 10)   #preceding spaces
## [1] "      2.72" "      7.39" "      20.1"
```

---

1. Pedantically, it accepts a path to a file, or a connection to a file, as returned by the file function.

```
formatC(powers_of_e, digits = 3, format = "e") #scientific formatting
## [1] "2.718e+00" "7.389e+00" "2.009e+01"
formatC(powers_of_e, digits = 3, flag = "+")   #precede +ve values with +
## [1] "+2.72" "+7.39" "+20.1"
```

R also provides slightly more general C-style formatting with the function sprintf. This works in the same way as sprintf in every other language: the first argument contains placeholders for string or number variables, and further arguments are substituted into those placeholders. Just remember that most numbers in R are floating-point values rather than integers.

The first argument to sprintf specifies a formatting string, with placeholders for other values. For example, %s denotes another string, %f and %e denote a floating-point number in fixed or scientific format, respectively, and %d represents an integer. Additional arguments specify the values to replace the placeholders. As with the paste function, shorter inputs are recycled to match the longest input:

```
sprintf("%s %d = %f", "Euler's constant to the power", pow, powers_of_e)
## [1] "Euler's constant to the power 1 = 2.718282"
## [2] "Euler's constant to the power 2 = 7.389056"
## [3] "Euler's constant to the power 3 = 20.085537"
sprintf("To three decimal places, e ^ %d = %.3f", pow, powers_of_e)
## [1] "To three decimal places, e ^ 1 = 2.718"
## [2] "To three decimal places, e ^ 2 = 7.389"
## [3] "To three decimal places, e ^ 3 = 20.086"
sprintf("In scientific notation, e ^ %d = %e", pow, powers_of_e)
## [1] "In scientific notation, e ^ 1 = 2.718282e+00"
## [2] "In scientific notation, e ^ 2 = 7.389056e+00"
## [3] "In scientific notation, e ^ 3 = 2.008554e+01"
```

Alternative syntaxes for formatting numbers are provided with the format and pretty Num functions. format just provides a slightly different syntax for formatting strings, and has similar usage to formatC. prettyNum, on the other hand, is best for pretty formatting of very big or very small numbers:

```
format(powers_of_e)
## [1] " 2.718" " 7.389" "20.086"
format(powers_of_e, digits = 3)                 #at least 3 sig figs
## [1] " 2.72" " 7.39" "20.09"
format(powers_of_e, digits = 3, trim = TRUE)    #remove leading zeros
## [1] "2.72"  "7.39"  "20.09"
```

```
format(powers_of_e, digits = 3, scientific = TRUE)

## [1] "2.72e+00" "7.39e+00" "2.01e+01"

prettyNum(
  c(1e10, 1e-20),
  big.mark     = ",",
  small.mark = " ",
  preserve.width = "individual",
  scientific = FALSE
)

## [1] "10,000,000,000"         "0.00000 00000 00000 00001"
```

Special Characters

There are some special characters that can be included in strings. For example, we can insert a tab character using \t. In the following example, we use cat rather than print, since print performs an extra conversion to turn \t from a tab character into a backslash and a "t." The argument fill = TRUE makes cat move the cursor to a new line after it is finished:

```
cat("foo\tbar", fill = TRUE)

## foo  bar
```

Moving the cursor to a new line is done by printing a newline character, \n (this is true on all platforms; don't try to use \r or \r\n for printing newlines to the R command line, since \r will just move the cursor to the start of the current line and overwrite what you have written):

```
cat("foo\nbar", fill = TRUE)

## foo
## bar
```

Null characters, \0, are used to terminate strings in R's internal code. It is an error to explicitly try to include them in a string (older versions of R will discard the contents of the string after the null character):

```
cat("foo\0bar", fill = TRUE)   #this throws an error
```

Backslash characters need to be doubled up so that they aren't mistaken for a special character. In this next example, the two input backslashes make just one backslash in the resultant string:

```
cat("foo\\bar", fill = TRUE)

## foo\bar
```

If we are using double quotes around our string, then double quote characters need to be preceded by a backslash to escape them. Similarly, if we use single quotes around our strings, then single quote characters need to be escaped:

```
cat("foo\"bar", fill = TRUE)
## foo"bar
cat('foo\'bar', fill = TRUE)
## foo'bar
```

In the converse case, including single quotes in double-quoted strings or double quotes inside single-quoted strings, we don't need an escape backslash:

```
cat("foo'bar", fill = TRUE)
## foo'bar
cat('foo"bar', fill = TRUE)
## foo"bar
```

We can make our computer beep by printing an alarm character, \a, though the function alarm will do this in a more readable way. This can be useful to add to the end of a long analysis to notify you that it's finished (as long as you aren't in an open-plan office):

```
cat("\a")
alarm()
```

Changing Case

Strings can be converted to contain all uppercase or all lowercase values using the functions toupper and tolower:

```
toupper("I'm Shouting")
## [1] "I'M SHOUTING"
tolower("I'm Whispering")
## [1] "i'm whispering"
```

Extracting Substrings

There are two related functions for extracting substrings: substring and substr. In most cases, it doesn't matter which you use, but they behave in slightly different ways when you pass vectors of different lengths as arguments. For substring, the output is as long as the longest input; for substr, the output is as long as the first input:

```
woodchuck <- c(
  "How much wood would a woodchuck chuck",
  "If a woodchuck could chuck wood?",
  "He would chuck, he would, as much as he could",
  "And chuck as much wood as a woodchuck would",
  "If a woodchuck could chuck wood."
)
```

```
substring(woodchuck, 1:6, 10)
```

```
## [1] "How much w" "f a woodc"  " would c"  " chuck "   " woodc"
## [6] "uch w"
```

```
substr(woodchuck, 1:6, 10)
```

```
## [1] "How much w" "f a woodc"  " would c"  " chuck "   " woodc"
```

Splitting Strings

The paste function and its friends combine strings together. strsplit does the opposite, breaking them apart at specified cut points. We can break the woodchuck tongue twister up into words by splitting it on the spaces. In the next example, fixed = TRUE means that the split argument is a fixed string rather than a regular expression:

```
strsplit(woodchuck, " ", fixed = TRUE)
```

```
## [[1]]
## [1] "How"       "much"      "wood"      "would"    "a"        "woodchuck"
## [7] "chuck"
##
## [[2]]
## [1] "If"        "a"         "woodchuck" "could"     "chuck"     "wood?"
##
## [[3]]
## [1] "He"        "would"   "chuck," "he"      "would," "as"      "much"
## [8] "as"        "he"      "could"
##
## [[4]]
## [1] "And"       "chuck"     "as"        "much"      "wood"      "as"
## [7] "a"         "woodchuck" "would"
##
## [[5]]
## [1] "If"        "a"         "woodchuck" "could"     "chuck"     "wood."
```

Notice that strsplit returns a list (not a character vector or matrix). This is because its result consists of character vectors of possibly different lengths. When you only pass a single string as an input, this fact is easy to overlook. Be careful!

In our example, the trailing commas on some words are a little annoying. It would be better to split on an optional comma followed by a space. This is easily specified using a regular expression. ? means "make the previous character optional":

```
strsplit(woodchuck, ",? ")
```

```
## [[1]]
## [1] "How"       "much"      "wood"      "would"    "a"        "woodchuck"
## [7] "chuck"
##
## [[2]]
## [1] "If"        "a"         "woodchuck" "could"     "chuck"     "wood?"
##
```

```
## [[3]]
## [1] "He"    "would" "chuck" "he"    "would" "as"    "much" "as"
## [9] "he"    "could"
##
## [[4]]
## [1] "And"    "chuck"    "as"      "much"    "wood"    "as"
## [7] "a"      "woodchuck" "would"
##
## [[5]]
## [1] "If"    "a"    "woodchuck" "could" "chuck" "wood."
```

File Paths

R has a working directory, which is the default place that files will be read from or written to. You can see its location with getwd and change it with setwd:

```
getwd()
```

```
## [1] "d:/workspace/LearningR"
```

```
setwd("c:/windows")
getwd()
```

```
## [1] "c:/windows"
```

Notice that the directory components of each path are separated by forward slashes, even though they are Windows pathnames. For portability, in R you can always specify paths with forward slashes, and the file handling functions will magically replace them with backslashes if the operating system needs them.

You can also specify a double backslash to denote Windows paths, but forward slashes are preferred:

```
"c:\\windows"          #remember to double up the slashes
"\\\\myserver\\mydir"  #UNC names need four slashes at the start
```

Alternatively, you can construct file paths from individual directory names using file.path. This automatically puts forward slashes between directory names. It's like a simpler, faster version of paste for paths:

```
file.path("c:", "Program Files", "R", "R-devel")
```

```
## [1] "c:/Program Files/R/R-devel"
```

```
R.home()    #same place: a shortcut to the R installation dir
```

```
## [1] "C:/PROGRA~1/R/R-devel"
```

Paths can be absolute (starting from a drive name or network share), or relative to the current working directory. In the latter case, . can be used for the current directory and .. can be used for the parent directory. ~ is shorthand for your user home directory. path.expand converts relative paths to absolute paths:

```
path.expand(".")
## [1] "."
path.expand("..")
## [1] ".."
path.expand("~")
## [1] "C:\\Users\\richie\\Documents"
```

basename returns the name of a file without the preceding directory location. Conversely, dirname returns the name of the directory that a file is in:

```
file_name <- "C:/Program Files/R/R-devel/bin/x64/RGui.exe"
basename(file_name)
## [1] "RGui.exe"
dirname(file_name)
## [1] "C:/Program Files/R/R-devel/bin/x64"
```

Factors

Factors are a special variable type for storing categorical variables. They sometimes behave like strings, and sometimes like integers.

Creating Factors

Whenever you create a data frame with a column of text data, R will assume by default that the text is categorical data and perform some conversion. The following example dataset contains the heights of 10 random adults:

```
(heights <- data.frame(
  height_cm = c(153, 181, 150, 172, 165, 149, 174, 169, 198, 163),
  gender = c(
    "female", "male", "female", "male", "male",
    "female", "female", "male", "male", "female"
  )
))
##     height_cm gender
## 1         153 female
## 2         181   male
## 3         150 female
## 4         172   male
## 5         165   male
## 6         149 female
## 7         174 female
## 8         169   male
## 9         198   male
## 10        163 female
```

By inspecting the class of the gender column we can see that it is not, as you may have expected, a character vector, but is in fact a factor:

```
class(heights$gender)
## [1] "factor"
```

Printing the column reveals a little more about the nature of this factor:

```
heights$gender
## [1] female male   female male   male   female female male   male   female
## Levels: female male
```

Each value in the factor is a string that is constrained to be either "female," "male," or missing. This constraint becomes obvious if we try to add a different string to the genders:

```
heights$gender[1] <- "Female"  #notice the capital "F"
## Warning: invalid factor level, NA generated
heights$gender
## [1] <NA>   male   female male   male   female female male   male   female
## Levels: female male
```

The choices "female" and "male" are called the *levels* of the factor and can be retrieved with the levels function:

```
levels(heights$gender)
## [1] "female" "male"
```

The number of these levels (equivalent to the length of the levels of the factor) can be retrieved with the nlevels function:

```
nlevels(heights$gender)
## [1] 2
```

Outside of their automatic creation inside data frames, you can create factors using the factor function. The first (and only compulsory) argument is a character vector:

```
gender_char <- c(
  "female", "male", "female", "male", "male",
  "female", "female", "male", "male", "female"
)
(gender_fac <- factor(gender_char))
## [1] female male   female male   male   female female male   male   female
## Levels: female male
```

Changing Factor Levels

We can change the order of the levels when the factor is created by specifying a `levels` argument:

```
factor(gender_char, levels = c("male", "female"))
## [1] female male   female male    male    female female male    male    female
## Levels: male female
```

If we want to change the order of the factor levels after creation, we again use the `factor` function, this time passing in the existing factor (rather than a character vector):

```
factor(gender_fac, levels = c("male", "female"))
## [1] female male   female male    male    female female male    male    female
## Levels: male female
```

 What we shouldn't do is directly change the levels using the `levels` function. This will relabel each level, changing data values, which is usually undesirable.

In the next example, directly setting the levels of the factor changes male data to female data, and female data to male data, which isn't what we want:

```
levels(gender_fac) <- c("male", "female")
gender_fac
## [1] male   female male   female female male    male    female female male
## Levels: male female
```

The `relevel` function is an alternative way of changing the order of factor levels. In this case, it just lets you specify which level comes first. As you might imagine, the use case for this function is rather niche—it can come in handy for regression models where you want to compare different categories to a reference category. Most of the time you will be better off calling `factor` if you want to set the levels:

```
relevel(gender_fac, "male")
## [1] male   female male   female female male    male    female female male
## Levels: male female
```

Dropping Factor Levels

In the process of cleaning datasets, you may end up removing all the data corresponding to a factor level. Consider this dataset of times to travel into work using different modes of transport:

```
getting_to_work <- data.frame(
  mode = c(
    "bike", "car", "bus", "car", "walk",
```

```
      "bike", "car", "bike", "car", "car"
  ),
    time_mins = c(25, 13, NA, 22, 65, 28, 15, 24, NA, 14)
  )
```

Not all the times have been recorded, so our first task is to remove the rows where time_mins is NA:

```
(getting_to_work <- subset(getting_to_work, !is.na(time_mins)))

##      mode time_mins
## 1   bike        25
## 2    car        13
## 4    car        22
## 5   walk        65
## 6   bike        28
## 7    car        15
## 8   bike        24
## 10   car        14
```

Looking at the mode column, there are now just three different values, but we still have the same *four* levels in the factor. We can see this with the unique function (the levels function will, of course, also tell us the levels of the factor):

```
unique(getting_to_work$mode)

## [1] bike car  walk
## Levels: bike bus car walk
```

If we want to drop the unused levels of the factor, we can use the droplevels function. This accepts either a factor or a data frame. In the latter case, it drops the unused levels in all the factors of the input. Since there is only one factor in our example data frame, the two lines of code in the next example are equivalent:

```
getting_to_work$mode <- droplevels(getting_to_work$mode)
getting_to_work <- droplevels(getting_to_work)
levels(getting_to_work$mode)

## [1] "bike" "car"  "walk"
```

Ordered Factors

Some factors have levels that are semantically greater than or less than other levels. This is common with multiple-choice survey questions. For example, the survey question "How happy are you?" could have the possible responses "depressed," "grumpy," "so-so," "cheery," and "ecstatic."[2] The resulting variable is categorical, so we can create a factor with the five choices as levels. Here we generate ten thousand random responses using the sample function:

---

2. This is sometimes called a Likert scale.

```
happy_choices <- c("depressed", "grumpy", "so-so", "cheery", "ecstatic")
happy_values <- sample(
  happy_choices,
  10000,
  replace = TRUE
)
happy_fac <- factor(happy_values, happy_choices)
head(happy_fac)
```

```
## [1] grumpy    depressed cheery    ecstatic  grumpy    grumpy
## Levels: depressed grumpy so-so cheery ecstatic
```

In this case, the five choices have a natural ordering to them: "grumpy" is happier than "depressed," "so-so" is happier than "grumpy," and so on. This means that it is better to store the responses in an *ordered factor*. We can do this using the ordered function (or by passing the argument ordered = TRUE to factor):

```
happy_ord <- ordered(happy_values, happy_choices)
head(happy_ord)
```

```
## [1] grumpy    depressed cheery    ecstatic  grumpy    grumpy
## Levels: depressed < grumpy < so-so < cheery < ecstatic
```

An ordered factor is a factor, but a normal factor isn't ordered:

```
is.factor(happy_ord)
```

```
## [1] TRUE
```

```
is.ordered(happy_fac)
```

```
## [1] FALSE
```

For most purposes, you don't need to worry about using ordered factors—you will only see a difference in some models—but they can be useful for analyzing survey data.

Converting Continuous Variables to Categorical

A useful way of summarizing a numeric variable is to count how many values fall into different "bins." The cut function cuts a numeric variable into pieces, returning a factor. It is commonly used with the table function to get counts of numbers in different groups. (The hist function, which draws histograms, provides an alternative way of doing this, as does count in the plyr package, which we will see later.)

In the next example, we randomly generate the ages of ten thousand workers (from 16 to 66, using a beta distribution) and put them in 10-year-wide groups:

```
ages <- 16 + 50 * rbeta(10000, 2, 3)
grouped_ages <- cut(ages, seq.int(16, 66, 10))
head(grouped_ages)
```

```
## [1] (26,36] (16,26] (26,36] (26,36] (26,36] (46,56]
## Levels: (16,26] (26,36] (36,46] (46,56] (56,66]
```

```
table(grouped_ages)
## grouped_ages
## (16,26] (26,36] (36,46] (46,56] (56,66]
##    1844    3339    3017    1533     267
```

In this case, the bulk of the workforce falls into the 26-to-36 and 36-to-46 categories (as a direct consequence of the shape of our beta distribution).

Notice that `ages` is a numeric variable and `grouped_ages` is a factor:

```
class(ages)
## [1] "numeric"
class(grouped_ages)
## [1] "factor"
```

Converting Categorical Variables to Continuous

The converse case of converting a factor into a numeric variable is most useful during data cleaning. If you have dirty data where numbers are mistyped, R may interpret them as strings and convert them to factors during the import process. In this next example, one of the numbers has a double decimal place. Import functions such as `read.table`, which we will look at in Chapter 12, would fail to parse such a string into numeric format, and default to making the column a character vector:

```
dirty <- data.frame(
  x = c("1.23", "4..56", "7.89")
)
```

To convert the x column to be numeric, the obvious solution is to call `as.numeric`. Unfortunately, it gives the wrong answer:

```
as.numeric(dirty$x)
## [1] 1 2 3
```

Calling `as.numeric` on a factor reveals the underlying integer codes that the factor uses to store its data. In general, a factor `f` can be reconstructed from `levels(f)[as.inte ger(f)]`.

To correctly convert the factor to numeric, we can first retrieve the values by converting the factor to a character vector. The second value is `NA` because `4..56` is not a genuine number:

```
as.numeric(as.character(dirty$x))
## Warning: NAs introduced by coercion
## [1] 1.23   NA 7.89
```

This is slightly inefficient, since repeated values have to be converted multiple times. As the FAQ on R (*http://bit.ly/13ZJ69Q*) notes, it is better to convert the factor's levels to be numeric, then reconstruct the factor as above:

```
as.numeric(levels(dirty$x))[as.integer(dirty$x)]
```

```
## Warning: NAs introduced by coercion
```

```
## [1] 1.23   NA 7.89
```

Since this is not entirely intuitive, if you want to do this task regularly, you can wrap it into a function for convenience:

```
factor_to_numeric <- function(f)
{
  as.numeric(levels(f))[as.integer(f)]
}
```

Generating Factor Levels

For balanced data, where there are an equal number of data points for each level, the gl function can be used to generate a factor. In its simplest form, the function takes an integer for the number of levels in the resultant factor, and another integer for how many times each level should be repeated. More commonly, you will want to set the names of the levels, which is achieved by passing a character vector to the labels argument. More complex level orderings, such as alternating values, can be created by also passing a length argument:

```
gl(3, 2)
```

```
## [1] 1 1 2 2 3 3
## Levels: 1 2 3
```

```
gl(3, 2, labels = c("placebo", "drug A", "drug B"))
```

```
## [1] placebo placebo drug A  drug A  drug B  drug B
## Levels: placebo drug A drug B
```

```
gl(3, 1, 6, labels = c("placebo", "drug A", "drug B")) #alternating
```

```
## [1] placebo drug A  drug B  placebo drug A  drug B
## Levels: placebo drug A drug B
```

Combining Factors

Where we have multiple categorical variables, it is sometimes useful to combine them into a single factor, where each level consists of the interactions of the individual variables:

```
treatment <- gl(3, 2, labels = c("placebo", "drug A", "drug B"))
gender <- gl(2, 1, 6, labels = c("female", "male"))
interaction(treatment, gender)
```

```
## [1] placebo.female placebo.male   drug A.female  drug A.male
## [5] drug B.female  drug B.male
## 6 Levels: placebo.female drug A.female drug B.female ... drug B.male
```

Summary

- You can combine strings together using `paste` and its derivatives.
- There are many functions for formatting numbers.
- Categorical data is stored in factors (or ordered factors).
- Each possible category in a factor is called a level.
- Continuous variables can be `cut` into categorical variables.

Test Your Knowledge: Quiz

Question 7-1
  Name as many functions as you can think of for formatting numbers.

Question 7-2
  How might you make your computer beep?

Question 7-3
  What are the classes of the two types of categorical variable?

Question 7-4
  What happens if you add a value to a factor that isn't one of the levels?

Question 7-5
  How do you convert a numeric variable to categorical?

Test Your Knowledge: Exercises

Exercise 7-1
  Display the value of *pi* to 16 significant digits. [5]

Exercise 7-2
  Split these strings into words, removing any commas or hyphens:

```
x <- c(
  "Swan swam over the pond, Swim swan swim!",
  "Swan swam back again - Well swum swan!"
)
```

  [5]

Exercise 7-3

For your role-playing game, each of your adventurer's character attributes is calculated as the sum of the scores from three six-sided dice rolls. To save arm-ache, you decide to use R to generate the scores. Here's a helper function to generate them:

```
#n specifies the number of scores to generate.
#It should be a natural number.
three_d6 <- function(n)
{
  random_numbers <- matrix(
    sample(6, 3 * n, replace = TRUE),
    nrow = 3
  )
  colSums(random_numbers)
}
```

Big scores give characters bonuses, and small scores give characters penalties, according to the following table:

Score	Bonus
3	−3
4, 5	−2
6 to 8	−1
9 to 12	0
13 to 15	+1
16, 17	+2
18	+3

Use the three_d6 function to generate 1,000 character attribute scores. Create a table of the number of scores with different levels of bonus. [15]

## CHAPTER 8

Flow Control and Loops

In R, as with other languages, there are many instances where we might want to conditionally execute code, or to repeatedly execute similar code.

The `if` and `switch` functions of R should be familiar if you have programmed in other languages, though the fact that they are functions may be new to you. Vectorized conditional execution via the `ifelse` function is also an R speciality.

We'll look at all of these in this chapter, as well as the three simplest loops (`for`, `while`, and `repeat`), which again should be reasonably familiar from other languages. Due to the vectorized nature of R, and some more aesthetic alternatives, these loops are less commonly used in R than you may expect.

Chapter Goals

After reading this chapter, you should:

- Be able to branch the flow of execution
- Be able to repeatedly execute code with loops

Flow Control

There are many occasions where you don't just want to execute one statement after another: you need to control the flow of execution. Typically this means that you only want to execute some code if a condition is fulfilled.

if and else

The simplest form of flow control is conditional execution using if. if takes a logical value (more precisely, a logical vector of length one) and executes the next statement only if that value is TRUE:

```
if(TRUE) message("It was true!")
## It was true!
if(FALSE) message("It wasn't true!")
```

Missing values aren't allowed to be passed to if; doing so throws an error:

```
if(NA) message("Who knows if it was true?")
## Error: missing value where TRUE/FALSE needed
```

Where you may have a missing value, you should test for it using is.na:

```
if(is.na(NA)) message("The value is missing!")
## The value is missing!
```

Of course, most of the time, you won't be passing the actual values TRUE or FALSE. Instead you'll be passing a variable or expression—if you knew that the statement was going to be executed in advance, you wouldn't need the if clause. In this next example, runif(1) generates one uniformly distributed random number between 0 and 1. If that value is more than 0.5, then the message is displayed:

```
if(runif(1) > 0.5) message("This message appears with a 50% chance.")
```

If you want to conditionally execute several statements, you can wrap them in curly braces:

```
x <- 3
if(x > 2)
{
  y <- 2 * x
  z <- 3 * y
}
```

For clarity of code, some style guides recommend always using curly braces, even if you only want to conditionally execute one statement.

The next step up in complexity from if is to include an else statement. Code that follows an else statement is executed if the if condition was FALSE:

```
if(FALSE)
{
  message("This won't execute...")
} else
{
  message("but this will.")
}

## but this will.
```

One important thing to remember is that the else statement must occur on the same line as the closing curly brace from the if clause. If you move it to the next line, you'll get an error:

```
if(FALSE)
{
  message("This won't execute...")
}
else
{
  message("and you'll get an error before you reach this.")
}
```

Multiple conditions can be defined by combining if and else repeatedly. Notice that if and else remain two separate words—there is an ifelse function but it means something slightly different, as we'll see in a moment:

```
(r <- round(rnorm(2), 1))

## [1] -0.1 -0.4

(x <- r[1] / r[2])

## [1] 0.25

if(is.nan(x))
{
  message("x is missing")
} else if(is.infinite(x))
{
  message("x is infinite")
} else if(x > 0)
{
  message("x is positive")
} else if(x < 0)
{
  message("x is negative")
} else
{
  message("x is zero")
}

## x is positive
```

R, unlike many languages, has a nifty trick that lets you reorder the code and do conditional assignment. In the next example, Re returns the real component of a complex number (Im returns the imaginary component):

```
x <- sqrt(-1 + 0i)
(reality <- if(Re(x) == 0) "real" else "imaginary")
## [1] "real"
```

Vectorized if

The standard if statement takes a single logical value. If you pass a logical vector with a length of more than one (don't do this!), then R will warn you that you've given multiple options, and only the first one will be used:

```
if(c(TRUE, FALSE)) message("two choices")

## Warning: the condition has length > 1 and only the first element will be
## used

## two choices
```

Since much of R is vectorized, you may not be surprised to learn that it also has vectorized flow control, in the form of the ifelse function. ifelse takes three arguments. The first is a logical vector of conditions. The second contains values that are returned when the first vector is TRUE. The third contains values that are returned when the first vector is FALSE. In the following example, rbinom generates random numbers from a binomial distribution to simulate a coin flip:

```
ifelse(rbinom(10, 1, 0.5), "Head", "Tail")

##  [1] "Head" "Head" "Head" "Tail" "Tail" "Head" "Head" "Head" "Tail" "Head"
```

ifelse can also accept vectors in the second and third arguments. These should be the same size as the first vector (if the vectors aren't the same size, then elements in the second and third arguments are recycled or ignored to make them the same size as the first):

```
(yn <- rep.int(c(TRUE, FALSE), 6))

##  [1]  TRUE FALSE  TRUE FALSE  TRUE FALSE  TRUE FALSE  TRUE FALSE  TRUE
## [12] FALSE

ifelse(yn, 1:3, -1:-12)

##  [1]   1  -2   3  -4   2  -6   1  -8   3 -10   2 -12
```

If there are missing values in the condition argument, then the corresponding values in the result will be missing:

```
yn[c(3, 6, 9, 12)] <- NA
ifelse(yn, 1:3, -1:-12)

##  [1]   1  -2  NA  -4   2  NA   1  -8  NA -10   2  NA
```

Multiple Selection

Code with many `else` statements can quickly become cumbersome to read. In such circumstances, prettier code can sometimes be achieved with a call to the `switch` function. The most common usage takes for its first argument an expression that returns a string, followed by several named arguments that provide results when the name matches the first argument. The names must match the first argument exactly (since R 2.11.0), and you can execute multiple expressions by enclosing them in curly braces:

```
(greek <- switch(
  "gamma",
  alpha = 1,
  beta  = sqrt(4),
  gamma =
  {
    a <- sin(pi / 3)
    4 * a ^ 2
  }
))
## [1] 3
```

If no names match, then `switch` (invisibly) returns `NULL`:

```
(greek <- switch(
  "delta",
  alpha = 1,
  beta  = sqrt(4),
  gamma =
  {
    a <- sin(pi / 3)
    4 * a ^ 2
  }
))
## NULL
```

For these circumstances, you can provide an unnamed argument that matches when nothing else does:

```
(greek <- switch(
  "delta",
  alpha = 1,
  beta  = sqrt(4),
  gamma =
  {
    a <- sin(pi / 3)
    4 * a ^ 2
  },
  4
))
## [1] 4
```

`switch` can also take a first argument that returns an integer. In this case the remaining arguments do not need names—the next argument is executed if the first argument resolves to 1, the argument after that is executed if the first argument resolves to 2, and so on:

```
switch(
  3,
  "first",
  "second",
  "third",
  "fourth"
)
## [1] "third"
```

As you may have noticed, no default argument is possible in this case. It's also rather cumbersome if you want to test for large integers, since you'll need to provide many arguments. Under those circumstances it is best to convert the first argument to a string and use the first syntax:

```
switch(
  as.character(2147483647),
  "2147483647" = "a big number",
  "another number"
)
## [1] "a big number"
```

Loops

There are three kinds of loops in R: `repeat`, `while`, and `for`. Although vectorization means that you don't need them as much in R as in other languages, they can still come in handy for repeatedly executing code.

repeat Loops

The easiest loop to master in R is `repeat`. All it does is execute the same code over and over until you tell it to stop. In other languages, it often goes by the name do `while`, or something similar. The following example[1] will execute until you press Escape, quit R, or the universe ends, whichever happens soonest:

---

1. If these examples make no sense, please watch the movie (*http://www.imdb.com/title/tt0107048*).

```
repeat
{
  message("Happy Groundhog Day!")
}
```

In general, we want our code to complete before the end of the universe, so it is possible to break out of the infinite loop by including a break statement. In the next example, sample returns one action in each iteration of the loop:

```
repeat
{
  message("Happy Groundhog Day!")
  action <- sample(
    c(
      "Learn French",
      "Make an ice statue",
      "Rob a bank",
      "Win heart of Andie McDowell"
    ),
    1
  )
  message("action = ", action)
  if(action == "Win heart of Andie McDowell") break
}
## Happy Groundhog Day!

## action = Rob a bank

## Happy Groundhog Day!

## action = Rob a bank

## Happy Groundhog Day!

## action = Rob a bank

## Happy Groundhog Day!

## action = Win heart of Andie McDowell
```

Sometimes, rather than breaking out of the loop we just want to skip the rest of the current iteration and start the next iteration:

```
repeat
{
  message("Happy Groundhog Day!")
  action <- sample(
    c(
      "Learn French",
      "Make an ice statue",
      "Rob a bank",
      "Win heart of Andie McDowell"
    ),
    1
  )
  if(action == "Rob a bank")
  {
    message("Quietly skipping to the next iteration")
    next
  }
  message("action = ", action)
  if(action == "Win heart of Andie McDowell") break
}
```

```
## Happy Groundhog Day!

## action = Learn French

## Happy Groundhog Day!

## Quietly skipping to the next iteration

## Happy Groundhog Day!

## Quietly skipping to the next iteration

## Happy Groundhog Day!

## action = Make an ice statue

## Happy Groundhog Day!

## action = Make an ice statue

## Happy Groundhog Day!

## Quietly skipping to the next iteration

## Happy Groundhog Day!

## action = Win heart of Andie McDowell
```

while Loops

while loops are like backward repeat loops. Rather than executing some code and then checking to see if the loop should end, they check first and then (maybe) execute. Since the check happens at the beginning, it is possible that the contents of the loop will *never* be executed (unlike in a repeat loop). The following example behaves similarly to the repeat example, except that if Andie McDowell's heart is won straightaway, then the *Groundhog Day* loop is completely avoided:

```r
action <- sample(
  c(
    "Learn French",
    "Make an ice statue",
    "Rob a bank",
    "Win heart of Andie McDowell"
  ),
  1
)
while(action != "Win heart of Andie McDowell")
{
  message("Happy Groundhog Day!")
  action <- sample(
    c(
      "Learn French",
      "Make an ice statue",
      "Rob a bank",
      "Win heart of Andie McDowell"
    ),
    1
  )
  message("action = ", action)
}
## Happy Groundhog Day!

## action = Make an ice statue

## Happy Groundhog Day!

## action = Learn French

## Happy Groundhog Day!

## action = Make an ice statue

## Happy Groundhog Day!

## action = Learn French

## Happy Groundhog Day!

## action = Make an ice statue

## Happy Groundhog Day!

## action = Win heart of Andie McDowell
```

With some fiddling, it is always possible to convert a repeat loop to a while loop or a while loop to a repeat loop, but usually the syntax is much cleaner one way or the other. If you know that the contents must execute at least once, use repeat; otherwise, use while.

for Loops

The third type of loop is to be used when you know exactly how many times you want the code to repeat. The `for` loop accepts an iterator variable and a vector. It repeats the loop, giving the iterator each element from the vector in turn. In the simplest case, the vector contains integers:

```
for(i in 1:5) message("i = ", i)
## i = 1
## i = 2
## i = 3
## i = 4
## i = 5
```

If you wish to execute multiple expressions, as with other loops they must be surrounded by curly braces:

```
for(i in 1:5)
{
  j <- i ^ 2
  message("j = ", j)
}
## j = 1
## j = 4
## j = 9
## j = 16
## j = 25
```

R's `for` loops are particularly flexible in that they are not limited to integers, or even numbers in the input. We can pass character vectors, logical vectors, or lists:

```
for(month in month.name)
{
  message("The month of ", month)
}
## The month of January
## The month of February
## The month of March
## The month of April
## The month of May
## The month of June
## The month of July
```

```
## The month of August

## The month of September

## The month of October

## The month of November

## The month of December

for(yn in c(TRUE, FALSE, NA))
{
  message("This statement is ", yn)
}

## This statement is TRUE

## This statement is FALSE

## This statement is NA

l <- list(
  pi,
  LETTERS[1:5],
  charToRaw("not as complicated as it looks"),
  list(
    TRUE
  )
)
for(i in l)
{
  print(i)
}

## [1] 3.142
## [1] "A" "B" "C" "D" "E"
##  [1] 6e 6f 74 20 61 73 20 63 6f 6d 70 6c 69 63 61 74 65 64 20 61 73 20 69
## [24] 74 20 6c 6f 6f 6b 73
## [[1]]
## [1] TRUE
```

Since for loops operate on each element of a vector, they provide a sort of "pretend vectorization." In fact, the vectorized operations in R will generally use some kind of for loop in internal C code. But be warned: R's for loops will almost always run much slower than their vectorized equivalents, often by an order of magnitude or two. This means that you should try to use the vectorization capabilities wherever possible.[2]

---

2. There is widespread agreement that if you write R code that looks like Fortran, you lose the right to complain that R is too slow.

Summary

- You can conditionally execute statements using `if` and `else`.
- The `ifelse` function is a vectorized equivalent of these.
- R has three kinds of loops: `repeat`, `while`, and `for`.

Test Your Knowledge: Quiz

Question 8-1
> What happens if you pass NA as a condition to `if`?

Question 8-2
> What happens if you pass NA as a condition to `ifelse`?

Question 8-3
> What types of variables can be passed as the first argument to the `switch` function?

Question 8-4
> How do you stop a `repeat` loop executing?

Question 8-5
> How do you jump to the next iteration of a loop?

Test Your Knowledge: Exercises

Exercise 8-1
> In the game of craps, the player (the "shooter") throws two six-sided dice. If the total is 2, 3, or 12, then the shooter loses. If the total is 7 or 11, she wins. If the total is any other score, then that score becomes the new target, known as the "point." Use this utility function to generate a craps score:

```
two_d6 <- function(n)
{
  random_numbers <- matrix(
    sample(6, 2 * n, replace = TRUE),
    nrow = 2
  )
  colSums(random_numbers)
}
```

> Write code that generates a craps score and assigns the following values to the variables `game_status` and `point`:

score	game_status	point
2, 3, 11	FALSE	NA
7, 11	TRUE	NA
4, 5, 6, 8, 9, 10	NA	Same as score

[10]

Exercise 8-2

If the shooter doesn't immediately win or lose, then he must keep rolling the dice until he scores the point value and wins, or scores a 7 and loses. Write code that checks to see if the game status is NA, and if so, repeatedly generates a craps score until either the point value is scored (set game_status to TRUE) or a 7 is scored (set game_status to FALSE). [15]

Exercise 8-3

This is the text for the famous "sea shells" tongue twister:

```
sea_shells <- c(
  "She", "sells", "sea", "shells", "by", "the", "seashore",
  "The", "shells", "she", "sells", "are", "surely", "seashells",
  "So", "if", "she", "sells", "shells", "on", "the", "seashore",
  "I'm", "sure", "she", "sells", "seashore", "shells"
)
```

Use the nchar function to calculate the number of letters in each word. Now loop over possible word lengths, displaying a message about which words have that length. For example, at length six, you should state that the words "shells" and "surely" have six letters. [10]

## CHAPTER 9

Advanced Looping

R's looping capability goes far beyond the three standard-issue loops seen in the last chapter. It gives you the ability to apply functions to each element of a vector, list, or array, so you can write pseudo-vectorized code where normal vectorization isn't possible. Other loops let you calculate summary statistics on chunks of data.

Chapter Goals

After reading this chapter, you should:

- Be able to apply a function to every element of a list or vector, or to every row or column of a matrix
- Be able to solve split-apply-combine problems
- Be able to use the `plyr` package

Replication

Cast your mind back to Chapter 4 and the `rep` function. `rep` repeats its input several times. Another related function, `replicate`, calls an expression several times. Mostly, they do exactly the same thing. The difference occurs when random number generation is involved. Pretend for a moment that the uniform random number generation function, `runif`, isn't vectorized. `rep` will repeat the same random number several times, but `replicate` gives a different number each time (for historical reasons, the order of the arguments is annoyingly back to front):

```
rep(runif(1), 5)
```

```
## [1] 0.04573 0.04573 0.04573 0.04573 0.04573
```

```
replicate(5, runif(1))
```

```
## [1] 0.5839 0.3689 0.1601 0.9176 0.5388
```

replicate comes into its own in more complicated examples: its main use is in Monte Carlo analyses, where you repeat an analysis a known number of times, and each iteration is independent of the others.

This next example estimates a person's time to commute to work via different methods of transport. It's a little bit complicated, but that's on purpose because that's when replicate is most useful.

The time_for_commute function uses sample to randomly pick a mode of transport (car, bus, train, or bike), then uses rnorm or rlnorm to find a normally or lognormally[1] distributed travel time (with parameters that depend upon the mode of transport):

```
time_for_commute <- function()
{
  #Choose a mode of transport for the day
  mode_of_transport <- sample(
    c("car", "bus", "train", "bike"),
    size = 1,
    prob = c(0.1, 0.2, 0.3, 0.4)
  )
  #Find the time to travel, depending upon mode of transport
  time <- switch(
    mode_of_transport,
    car   = rlnorm(1, log(30), 0.5),
    bus   = rlnorm(1, log(40), 0.5),
    train = rnorm(1, 30, 10),
    bike  = rnorm(1, 60, 5)
  )
  names(time) <- mode_of_transport
  time
}
```

The presence of a switch statement makes this function very hard to vectorize. That means that to find the distribution of commuting times, we need to repeatedly call time_for_commute to generate data for each day. replicate gives us instant vectorization:

```
replicate(5, time_for_commute())
```

```
##  bike    car train    bus  bike
## 66.22 35.98 27.30 39.40 53.81
```

---

1. Lognormal distributions occasionally throw out very big numbers, thus approximating rush hour gridlock.

Looping Over Lists

By now, you should have noticed that an awful lot of R is vectorized. In fact, your default stance should be to write vectorized code. It's often cleaner to read, and invariably gives you performance benefits when compared to a loop. In some cases, though, trying to achieve vectorization means contorting your code in unnatural ways. In those cases, the apply family of functions can give you pretend vectorization,[2] without the pain.

The simplest and most commonly used family member is lapply, short for "list apply." lapply takes a list and a function as inputs, applies the function to each element of the list in turn, and returns another list of results. Recall our prime factorization list from Chapter 5:

```
prime_factors <- list(
  two   = 2,
  three = 3,
  four  = c(2, 2),
  five  = 5,
  six   = c(2, 3),
  seven = 7,
  eight = c(2, 2, 2),
  nine  = c(3, 3),
  ten   = c(2, 5)
)
head(prime_factors)

## $two
## [1] 2
##
## $three
## [1] 3
##
## $four
## [1] 2 2
##
## $five
## [1] 5
##
## $six
## [1] 2 3
##
## $seven
## [1] 7
```

Trying to find the unique value in each list element is difficult to do in a vectorized way. We could write a for loop to examine each element, but that's a little bit clunky:

---

2. Since the vectorization happens at the R level rather than by calling internal C code, you don't get the performance benefits of the vectorization, only more readable code.

```
unique_primes <- vector("list", length(prime_factors))
for(i in seq_along(prime_factors))
{
  unique_primes[[i]] <- unique(prime_factors[[i]])
}
names(unique_primes) <- names(prime_factors)
unique_primes
```

```
## $two
## [1] 2
##
## $three
## [1] 3
##
## $four
## [1] 2
##
## $five
## [1] 5
##
## $six
## [1] 2 3
##
## $seven
## [1] 7
##
## $eight
## [1] 2
##
## $nine
## [1] 3
##
## $ten
## [1] 2 5
```

lapply makes this so much easier, eliminating the nasty boilerplate code for worrying about lengths and names:

```
lapply(prime_factors, unique)
```

```
## $two
## [1] 2
##
## $three
## [1] 3
##
## $four
## [1] 2
##
## $five
## [1] 5
##
## $six
```

```
## [1] 2 3
##
## $seven
## [1] 7
##
## $eight
## [1] 2
##
## $nine
## [1] 3
##
## $ten
## [1] 2 5
```

When the return value from the function is the same size each time, and you know what that size is, you can use a variant of `lapply` called `vapply`. `vapply` stands for "list apply that returns a vector." As before, you pass it a list and a function, but `vapply` takes a third argument that is a template for the return values. Rather than returning a list, it simplifies the result to be a vector or an array:

```
vapply(prime_factors, length, numeric(1))
##  two three  four  five   six seven eight  nine   ten
##    1     1     2     1     2     1     3     2     2
```

If the output does not fit the template, then `vapply` will throw an error. This makes it less flexible than `lapply`, since the output must be the same size for each element and must be known in advance.

There is another function that lies in between `lapply` and `vapply`: namely `sapply`, which stands for "simplifying list apply." Like the two other functions, `sapply` takes a list and a function as inputs. It does not need a template, but will try to simplify the result to an appropriate vector or array if it can:

```
sapply(prime_factors, unique)  #returns a list
## $two
## [1] 2
##
## $three
## [1] 3
##
## $four
## [1] 2
##
## $five
## [1] 5
##
## $six
## [1] 2 3
##
## $seven
```

```
## [1] 7
##
## $eight
## [1] 2
##
## $nine
## [1] 3
##
## $ten
## [1] 2 5
```

```
sapply(prime_factors, length)  #returns a vector
```

```
##  two three  four  five   six seven eight  nine   ten
##    1     1     2     1     2     1     3     2     2
```

```
sapply(prime_factors, summary) #returns an array
```

```
##          two three four five  six seven eight nine  ten
## Min.       2     3    2    5 2.00     7     2    3 2.00
## 1st Qu.    2     3    2    5 2.25     7     2    3 2.75
## Median     2     3    2    5 2.50     7     2    3 3.50
## Mean       2     3    2    5 2.50     7     2    3 3.50
## 3rd Qu.    2     3    2    5 2.75     7     2    3 4.25
## Max.       2     3    2    5 3.00     7     2    3 5.00
```

For interactive use, this is wonderful because you usually automatically get the result in the form that you want. This function does require some care if you aren't sure about what your inputs might be, though, since the result is sometimes a list and sometimes a vector. This can trip you up in some subtle ways. Our previous length example returned a vector, but look what happens when you pass it an empty list:

```
sapply(list(), length)
```

```
## list()
```

If the input list has length zero, then sapply always returns a list, regardless of the function that is passed. So if your data could be empty, and you know the return value, it is safer to use vapply:

```
vapply(list(), length, numeric(1))
```

```
## numeric(0)
```

Although these functions are primarily designed for use with lists, they can also accept vector inputs. In this case, the function is applied to each element of the vector in turn. The source function is used to read and evaluate the contents of an R file. (That is, you can use it to run an R script.) Unfortunately it isn't vectorized, so if we wanted to run all the R scripts in a directory, then we need to wrap the directory in a call to lapply.

In this next example, dir returns the names of files in a given directory, defaulting to the current working directory. (Recall that you can find this with getwd.) The argument pattern = "\\.R$" means "only return filenames that end with .R":

```
r_files <- dir(pattern = "\\.R$")
lapply(r_files, source)
```

You may have noticed that in all of our examples, the functions passed to lapply, vapply, and sapply have taken just one argument. There is a limitation in these functions in that you can only pass one *vectorized* argument (more on how to circumvent that later), but you can pass other scalar arguments to the function. To do this, just pass in named arguments to the lapply (or sapply or vapply) call, and they will be passed to the inner function. For example, if rep.int takes two arguments, but the times argument is allowed to be a single (scalar) number, you'd type:

```
complemented <- c(2, 3, 6, 18)          #See http://oeis.org/A000614
lapply(complemented, rep.int, times = 4)

## [[1]]
## [1] 2 2 2 2
##
## [[2]]
## [1] 3 3 3 3
##
## [[3]]
## [1] 6 6 6 6
##
## [[4]]
## [1] 18 18 18 18
```

What if the vector argument isn't the first one? In that case, we have to create our own function to wrap the function that we really wanted to call. You can do this on a separate line, but it is common to include the function definition within the call to lapply:

```
rep4x <- function(x) rep.int(4, times = x)
lapply(complemented, rep4x)

## [[1]]
## [1] 4 4
##
## [[2]]
## [1] 4 4 4
##
## [[3]]
## [1] 4 4 4 4 4 4
##
## [[4]]
## [1] 4 4 4 4 4 4 4 4 4 4 4 4 4 4 4 4 4 4
```

This last code chunk can be made a little simpler by passing an anonymous function to lapply. This is the trick we saw in Chapter 5, where we don't bother with a separate assignment line and just pass the function to lapply without giving it a name:

```
lapply(complemented, function(x) rep.int(4, times = x))
## [[1]]
## [1] 4 4
##
## [[2]]
## [1] 4 4 4
##
## [[3]]
## [1] 4 4 4 4 4 4
##
## [[4]]
##  [1] 4 4 4 4 4 4 4 4 4 4 4 4 4 4 4 4 4 4
```

Very, very occasionally, you may want to loop over every variable in an environment, rather than in a list. There is a dedicated function, eapply, for this, though in recent versions of R you can also use lapply:

```
env <- new.env()
env$molien <- c(1, 0, 1, 0, 1, 1, 2, 1, 3) #See http://oeis.org/A008584
env$larry <- c("Really", "leery", "rarely", "Larry")
eapply(env, length)
## $molien
## [1] 9
##
## $larry
## [1] 4

lapply(env, length) #same
## $molien
## [1] 9
##
## $larry
## [1] 4
```

rapply is a recursive version of lapply that allows you to loop over nested lists. This is a niche requirement, and code is often simpler if you flatten the data first using unlist.

Looping Over Arrays

lapply, and its friends vapply and sapply, can be used on matrices and arrays, but their behavior often isn't what we want. The three functions treat the matrices and arrays as though they were vectors, applying the target function to each element one at a time (moving down columns). More commonly, when we want to apply a function to an array, we want to apply it by row or by column. This next example uses the matlab package, which gives some functionality ported from the rival language.

To run the next example, you first need to install the matlab package:

```
install.packages("matlab")

library(matlab)

## Attaching package: 'matlab'

## The following object is masked from 'package:stats':
##
## reshape

## The following object is masked from 'package:utils':
##
## find, fix

## The following object is masked from 'package:base':
##
## sum
```

 When you load the matlab package, it overrides some functions in the base, stats, and utils packages to make them behave like their MATLAB counterparts. After these examples that use the matlab package, you may wish to restore the usual behavior by unloading the package. Call detach("package:matlab") to do this.

The magic function creates a magic square—an n-by-n square matrix of the numbers from 1 to n^2, where each row and each column has the same total:

```
(magic4 <- magic(4))

##      [,1] [,2] [,3] [,4]
## [1,]   16    2    3   13
## [2,]    5   11   10    8
## [3,]    9    7    6   12
## [4,]    4   14   15    1
```

A classic problem requiring us to apply a function by row is calculating the row totals. This can be achieved using the rowSums function that we saw briefly in Chapter 5:

```
rowSums(magic4)

## [1] 34 34 34 34
```

But what if we want to calculate a different statistic for each row? It would be cumbersome to try to provide a function for every such possibility.[3] The apply function provides the row/column-wise equivalent of lapply, taking a matrix, a dimension number, and a function as arguments. The dimension number is 1 for "apply the function across each row," or 2 for "apply the function down each column" (or bigger numbers for higher-dimensional arrays):

---

3. Though the matrixStats package tries to do exactly that.

```
apply(magic4, 1, sum)        #same as rowSums
## [1] 34 34 34 34

apply(magic4, 1, toString)
## [1] "16, 2, 3, 13" "5, 11, 10, 8" "9, 7, 6, 12"  "4, 14, 15, 1"

apply(magic4, 2, toString)
## [1] "16, 5, 9, 4"  "2, 11, 7, 14" "3, 10, 6, 15" "13, 8, 12, 1"
```

apply can also be used on data frames, though the mixed-data-type nature means that this is less common (for example, you can't sensibly calculate a sum or a product when there are character columns):

```
(baldwins <- data.frame(
  name           = c("Alec", "Daniel", "Billy", "Stephen"),
  date_of_birth  = c(
    "1958-Apr-03", "1960-Oct-05", "1963-Feb-21", "1966-May-12"
  ),
  n_spouses      = c(2, 3, 1, 1),
  n_children     = c(1, 5, 3, 2),
  stringsAsFactors = FALSE
))

##       name date_of_birth n_spouses n_children
## 1     Alec   1958-Apr-03         2          1
## 2   Daniel   1960-Oct-05         3          5
## 3    Billy   1963-Feb-21         1          3
## 4  Stephen   1966-May-12         1          2

apply(baldwins, 1, toString)
## [1] "Alec, 1958-Apr-03, 2, 1"    "Daniel, 1960-Oct-05, 3, 5"
## [3] "Billy, 1963-Feb-21, 1, 3"   "Stephen, 1966-May-12, 1, 2"

apply(baldwins, 2, toString)
##                                                     name
##                          "Alec, Daniel, Billy, Stephen"
##                                            date_of_birth
## "1958-Apr-03, 1960-Oct-05, 1963-Feb-21, 1966-May-12"
##                                                n_spouses
##                                          "2, 3, 1, 1"
##                                               n_children
##                                          "1, 5, 3, 2"
```

When applied to a data frame by column, apply behaves identically to sapply (remember that data frames can be thought of as nonnested lists where the elements are of the same length):

```
sapply(baldwins, toString)
##                                                     name
##                          "Alec, Daniel, Billy, Stephen"
##                                            date_of_birth
```

```
## "1958-Apr-03, 1960-Oct-05, 1963-Feb-21, 1966-May-12"
##                                              n_spouses
##                                            "2, 3, 1, 1"
##                                             n_children
##                                            "1, 5, 3, 2"
```

Of course, simply printing a dataset in different forms isn't that interesting. Using `sapply` combined with `range`, on the other hand, is a great way to quickly determine the extent of your data:

```
sapply(baldwins, range)
```

```
##         name     date_of_birth n_spouses n_children
## [1,] "Alec"     "1958-Apr-03" "1"       "1"
## [2,] "Stephen"  "1966-May-12" "3"       "5"
```

Multiple-Input Apply

One of the drawbacks of `lapply` is that it only accepts a single vector to loop over. Another is that inside the function that is called on each element, you don't have access to the name of that element.

The function `mapply`, short for "multiple argument list apply," lets you pass in as many vectors as you like, solving the first problem. A common usage is to pass in a list in one argument and the names of that list in another, solving the second problem. One little annoyance is that in order to accommodate an arbitrary number of vector arguments, the order of the arguments has been changed. For `mapply`, the function is passed as the first argument:

```
msg <- function(name, factors)
{
  ifelse(
    length(factors) == 1,
    paste(name, "is prime"),
    paste(name, "has factors", toString(factors))
  )
}
mapply(msg, names(prime_factors), prime_factors)
```

```
##                          two                        three
##               "two is prime"               "three is prime"
##                         four                         five
##      "four has factors 2, 2"               "five is prime"
##                          six                        seven
##       "six has factors 2, 3"               "seven is prime"
##                        eight                         nine
## "eight has factors 2, 2, 2"     "nine has factors 3, 3"
##                          ten
##       "ten has factors 2, 5"
```

By default, `mapply` behaves in the same way as `sapply`, simplifying the output if it thinks it can. You can turn this behavior off (so it behaves more like `lapply`) by passing the argument `SIMPLIFY = FALSE`.

Instant Vectorization

The function `Vectorize` is a wrapper to `mapply` that takes a function that usually accepts a scalar input, and returns a new function that accepts vectors. This next function is not vectorized because of its use of `switch`, which requires a scalar input:

```
baby_gender_report <- function(gender)
{
  switch(
    gender,
    male   = "It's a boy!",
    female = "It's a girl!",
    "Um..."
  )
}
```

If we pass a vector into the function, it will throw an error:

```
genders <- c("male", "female", "other")
baby_gender_report(genders)
```

While it is theoretically possible to do a complete rewrite of a function that is inherently vectorized, it is easier to use the `Vectorize` function:

```
vectorized_baby_gender_report <- Vectorize(baby_gender_report)
vectorized_baby_gender_report(genders)
```

```
##          male          female          other
##   "It's a boy!" "It's a girl!"        "Um..."
```

Split-Apply-Combine

A really common problem when investigating data is how to calculate some statistic on a variable that has been split into groups. Here are some scores on the classic road safety awareness computer game, Frogger:

```
(frogger_scores <- data.frame(
  player = rep(c("Tom", "Dick", "Harry"), times = c(2, 5, 3)),
  score  = round(rlnorm(10, 8), -1)
))
```

```
##      player score
## 1      Tom  2250
## 2      Tom  1510
## 3     Dick  1700
## 4     Dick   410
## 5     Dick  3720
```

```
## 6      Dick  1510
## 7      Dick  4500
## 8     Harry  2160
## 9     Harry  5070
## 10    Harry  2930
```

If we want to calculate the mean score for each player, then there are three steps. First, we *split* the dataset by player:

```
(scores_by_player <- with(
  frogger_scores,
  split(score, player)
))
## $Dick
## [1] 1700  410 3720 1510 4500
##
## $Harry
## [1] 2160 5070 2930
##
## $Tom
## [1] 2250 1510
```

Next we *apply* the (mean) function to each element:

```
(list_of_means_by_player <- lapply(scores_by_player, mean))
## $Dick
## [1] 2368
##
## $Harry
## [1] 3387
##
## $Tom
## [1] 1880
```

Finally, we *combine* the result into a single vector:

```
(mean_by_player <- unlist(list_of_means_by_player))
##  Dick Harry   Tom
##  2368  3387  1880
```

The last two steps can be condensed into one by using vapply or sapply, but split-apply-combine is such a common task that we need something easier. That something is the tapply function, which performs all three steps in one go:

```
with(frogger_scores, tapply(score, player, mean))
##  Dick Harry   Tom
##  2368  3387  1880
```

There are a few other wrapper functions to tapply, namely by and aggregate. They perform the same function with a slightly different interface.

SQL fans may note that split-apply-combine is the same as a GROUP BY operation.

The plyr Package

The *apply family of functions are mostly wonderful, but they have three drawbacks that stop them being as easy to use as they could be. Firstly, the names are a bit obscure. The "l" in lapply for lists makes sense, but after using R for nine years, I still don't know what the "t" in tapply stands for.

Secondly, the arguments aren't entirely consistent. Most of the functions take a data object first and a function argument second, but mapply swaps the order, and tapply takes the function for its third argument. The data argument is sometimes X and sometimes object, and the simplification argument is sometimes simplify and sometimes SIMPLIFY.

Thirdly, the form of the output isn't as controllable as it could be. Getting your results as a data frame—or discarding the result—takes a little bit of effort.

This is where the plyr package comes in handy. The package contains a set of functions named **ply, where the blanks (asterisks) denote the form of the input and output, respectively. So, llply takes a list input, applies a function to each element, and returns a list, making it a drop-in replacement for lapply:

```
library(plyr)
llply(prime_factors, unique)

## $two
## [1] 2
##
## $three
## [1] 3
##
## $four
## [1] 2
##
## $five
## [1] 5
##
## $six
## [1] 2 3
##
## $seven
## [1] 7
##
## $eight
## [1] 2
```

```
##
## $nine
## [1] 3
##
## $ten
## [1] 2 5
```

laply takes a list and returns an array, mimicking sapply. In the case of an empty input, it does the smart thing and returns an empty logical vector (unlike sapply, which returns an empty list):

```
laply(prime_factors, length)

## [1] 1 1 2 1 2 1 3 2 2

laply(list(), length)

## logical(0)
```

raply replaces replicate (*not* rapply!), but there are also rlply and rdply functions that let you return the result in list or data frame form, and an r_ply function that discards the result (useful for drawing plots):

```
raply(5, runif(1))  #array output

## [1] 0.009415 0.226514 0.133015 0.698586 0.112846

rlply(5, runif(1))  #list output

## [[1]]
## [1] 0.6646
##
## [[2]]
## [1] 0.2304
##
## [[3]]
## [1] 0.613
##
## [[4]]
## [1] 0.5532
##
## [[5]]
## [1] 0.3654

rdply(5, runif(1))  #data frame output

##   .n     V1
## 1  1 0.9068
## 2  2 0.0654
## 3  3 0.3788
## 4  4 0.5086
## 5  5 0.3502

r_ply(5, runif(1))  #discarded output

## NULL
```

Perhaps the most commonly used function in `plyr` is `ddply`, which takes data frames as inputs and outputs and can be used as a replacement for `tapply`. Its big strength is that it makes it easy to make calculations on several columns at once. Let's add a `level` column to the Frogger dataset, denoting the level the player reached in the game:

```
frogger_scores$level <- floor(log(frogger_scores$score))
```

There are several different ways of calling `ddply`. All methods take a data frame, the name of the column(s) to split by, and the function to apply to each piece. The column is passed without quotes, but wrapped in a call to the . function.

For the function, you can either use `colwise` to tell `ddply` to call the function on every column (that you didn't mention in the second argument), or use `summarize` and specify manipulations of specific columns:

```
ddply(
  frogger_scores,
  .(player),
  colwise(mean) #call mean on every column except player
)
##    player score level
## 1   Dick  2368 7.200
## 2  Harry  3387 7.333
## 3    Tom  1880 7.000
ddply(
  frogger_scores,
  .(player),
  summarize,
  mean_score = mean(score), #call mean on score
  max_level  = max(level)   #... and max on level
)
##    player mean_score max_level
## 1   Dick       2368         8
## 2  Harry       3387         8
## 3    Tom       1880         7
```

`colwise` is quicker to specify, but you have to do the same thing with each column, whereas `summarize` is more flexible but requires more typing.

There is no direct replacement for `mapply`, though the `m*ply` functions allow looping with multiple arguments. Likewise, there is no replacement for `vapply` or `rapply`.

Summary

- The `apply` family of functions provide cleaner code for looping.
- Split-apply-combine problems, where you manipulate data split into groups, are really common.
- The `plyr` package is a syntactically cleaner replacement for many apply functions.

Test Your Knowledge: Quiz

Question 9-1

    Name as many members of the `apply` family of functions as you can.

Question 9-2

    What is the difference between `lapply`, `vapply`, and `sapply`?

Question 9-3

    How might you loop over tree-like data?

Question 9-4

    Given some height data, how might you calculate a mean height by gender?

Question 9-5

    In the `plyr` package, what do the asterisks mean in a name like `**ply`?

Test Your Knowledge: Exercises

Exercise 9-1

    Loop over the list of children in the celebrity Wayans family. How many children does each of the first generation of Wayanses have?

```
wayans <- list(
  "Dwayne Kim" = list(),
  "Keenen Ivory" = list(
    "Jolie Ivory Imani",
    "Nala",
    "Keenen Ivory Jr",
    "Bella",
    "Daphne Ivory"
  ),
  Damon = list(
    "Damon Jr",
    "Michael",
    "Cara Mia",
    "Kyla"
  ),
```

```
      Kim = list(),
      Shawn = list(
        "Laila",
        "Illia",
        "Marlon"
      ),
      Marlon = list(
        "Shawn Howell",
        "Arnai Zachary"
      ),
      Nadia = list(),
      Elvira = list(
        "Damien",
        "Chaunté"
      ),
      Diedre = list(
        "Craig",
        "Gregg",
        "Summer",
        "Justin",
        "Jamel"
      ),
      Vonnie = list()
    )
```

[5]

Exercise 9-2

state.x77 is a dataset that is supplied with R. It contains information about the population, income, and other factors for each US state. You can see its values by typing its name, just as you would with datasets that you create yourself:

```
state.x77
```

1. Inspect the dataset using the method that you learned in Chapter 3.

2. Find the mean and standard deviation of each column.

   [10]

Exercise 9-3

Recall the time_for_commute function from earlier in the chapter. Calculate the 75th-percentile commute time by mode of transport:

```
commute_times <- replicate(1000, time_for_commute())
commute_data <- data.frame(
  time = commute_times,
  mode = names(commute_times)
)
```

[5]

## CHAPTER 10

Packages

R is not limited to the code provided by the R Core Team. It is very much a community effort, and there are thousands of add-on packages available to extend it. The majority of R packages are currently installed in an online repository called CRAN (the Comprehensive R Archive Network[1]), which is maintained by the R Core Team. Installing and using these add-on packages is an important part of the R experience.

We've just seen the `plyr` package for advanced looping. Throughout the rest of the book, we'll see many more common packages: `lubridate` for date and time manipulation, `xlsx` for importing Excel files, `reshape2` for manipulating the shape of data frames, `ggplot2` for plotting, and dozens of others.[2]

Chapter Goals

After reading this chapter, you should:

- Be able to load packages that are installed on your machine
- Know how to install new packages from local files and via the Internet
- Understand how to manage the packages on your machine

---

1. Named after CPAN, the Comprehensive Perl Archive Network.

2. You can find a list of the most popular packages on the R-statistics blog (*http://www.r-statistics.com/2013/06/top-100-r-packages-for-2013-jan-may*).

Loading Packages

To load a package that is already installed on your machine, you call the `library` function. It is widely agreed that calling this function `library` was a mistake, and that calling it `load_package` would have saved a lot of confusion, but the function has existed long enough that it is too late to change it now. To clarify the terminology, a *package* is a collection of R functions and datasets, and a *library* is a folder on your machine that stores the files for a package.[3]

If you have a standard version of R—that is, you haven't built some custom version from the source code—the `lattice` package should be installed, but it won't automatically be loaded. We can load it with the `library` function:

```
library(lattice)
```

We can now use all the functions provided by `lattice`. For example, Figure 10-1 displays a fancy dot plot of the famous Immer's barley dataset:

```
dotplot(
  variety ~ yield | site,
  data  = barley,
  groups = year
)
```

The `lattice` package is covered in detail in Chapter 14.

Notice that the name of the package is passed to `library` without being enclosed in quotes. If you want to programmatically pass the name of the package to `library`, then you can set the argument `character.only = TRUE`. This is mildly useful if you have a lot of packages to load:

```
pkgs <- c("lattice", "utils", "rpart")
for(pkg in pkgs)
{
  library(pkg, character.only = TRUE)
}
```

3. Be warned that some people on the *R-help* mailing list consider confusing the two pieces of terminology a capital offense.

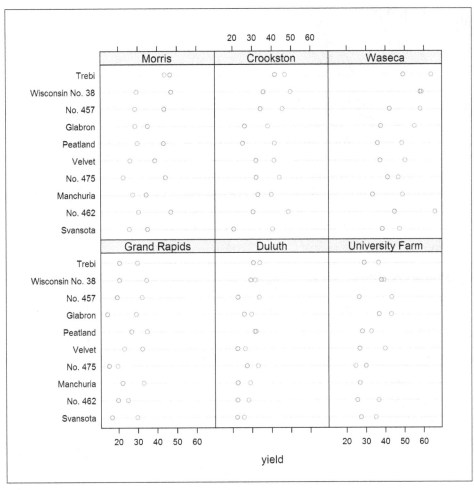

Figure 10-1. A dot plot of Immer's barley data using lattice

If you use library to try to load a package that isn't installed, then it will throw an error. Sometimes you might want to handle the situation differently, in which case the require function provides an alternative. Like library, require loads a package, but rather than throwing an error it returns TRUE or FALSE, depending upon whether or not the package was successfully loaded:

```
if(!require(apackagethatmightnotbeinstalled))
{
  warning("The package 'apackagethatmightnotbeinstalled' is not available.")
  #perhaps try to download it
  #...
}
```

The Search Path

You can see the packages that are loaded with the `search` function:

```
search()
## [1] ".GlobalEnv"        "package:stats"    "package:graphics"
## [4] "package:grDevices" "package:utils"    "package:datasets"
## [7] "package:methods"   "Autoloads"        "package:base"
```

This list shows the order of places that R will look to try to find a variable. The global environment always comes first, followed by the most recently loaded packages. The last two values are always a special environment called `Autoloads`, and the base package. If you define a variable called `var` in the global environment, R will find that before it finds the usual variance function in the `stats` package, because the global environment comes first in the search list. If you create any environments (see Chapter 6), they will also appear on the search path.

Libraries and Installed Packages

The function `installed.packages` returns a data frame with information about all the packages that R knows about on your machine. If you've been using R for a while, this can easily be several hundred packages, so it is often best to view the results away from the console:

```
View(installed.packages())
```

`installed.packages` gives you information about which version of each package is installed, where it lives on your hard drive, and which other packages it depends upon, amongst other things. The `LibPath` column that provides the file location of the package tells you the library that contains the package. At this point, you may be wondering how R decides which folders are considered libraries.

 The following explanation is a little bit technical, so don't worry about remembering the minutiae of how R finds its packages. This information can save you administration effort when you choose to upgrade R, or when you have problems with packages loading, but it isn't required for day-to-day use of R.

The packages that come with the R install (`base`, `stats`, and nearly 30 others) are stored in the *library* subdirectory of wherever you installed R. You can retrieve the location of this with:

```
R.home("library")   #or
## [1] "C:/PROGRA~1/R/R-devel/library"
```

```
.Library
## [1] "C:/PROGRA~1/R/R-devel/library"
```

You also get a user library for installing packages that will only be accessible by you. (This is useful if you install on a family PC and don't want your six-year-old to update packages and break compatibility in your code.) The location is OS dependent. Under Windows, for R version *x.y.z*, it is in the *R/win-library/x.y* subfolder of the home directory, where the home directory can be found via:

```
path.expand("~")    #or
## [1] "C:\\Users\\richie\\Documents"
Sys.getenv("HOME")
## [1] "C:\\Users\\richie\\Documents"
```

Under Linux, the folder is similarly located in the *R/R.version$platform-library/x.y* subfolder of the home directory. R.version$platform will typically return a string like "i686-pc-linux-gnu," and the home directory is found in the same way as under Windows. Under Mac OS X, it is found in *Library/R/x.y/library*.

One problem with the default setup of library locations is that when you upgrade R, you need to reinstall all your packages. This is the safest behavior, since different versions of R will often need different versions of packages. In practice, on a development machine the convenience of not having to reinstall packages often outweighs versioning worries.[4] To make life easier for yourself, it's a very good idea to create your own library that can be used by all versions of R. The simplest way of doing this is to define an environment variable named R_LIBS that contains a path[5] to your desired library location. Although you can define environment variables programmatically with R, they are only available to R, and only for the rest of the session—define them from within your operating system instead.

You can see a character vector of all the libraries that R knows about using the .lib Paths function:

```
.libPaths()
## [1] "D:/R/library"
## [2] "C:/Program Files/R/R-devel/library"
```

The first value in this vector is the most important, as this is where packages will be installed by default.

---

4. If you are deploying R code as part of an application, then the situation is different. Compatibility trumps convenience in that case.

5. You can have multiple locations separated by semicolons, but this is usually overkill.

Installing Packages

Factory-fresh installs of R are set up to access the CRAN package repository (via a mirror —you'll be prompted to pick the one nearest to you), and CRANextra if you are running Windows. CRANextra contains a handful of packages that need special attention to build under Windows, and cannot be hosted on the usual CRAN servers. To access additional repositories, type `setRepositories()` and select the repositories that you want. Figure 10-2 shows the available options.

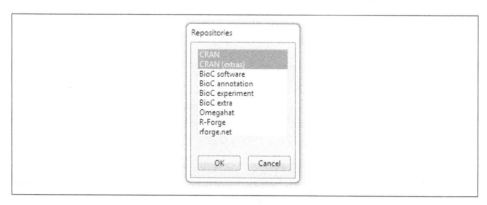

Figure 10-2. List of available package repositories

Bioconductor contains packages related to genomics and molecular biology, while R-Forge and RForge.net mostly contain development versions of packages that eventually appear on CRAN. You can see information about all the packages that are available in the repositories that you have set using `available.packages` (be warned—there are thousands, so this takes several seconds to run):

```
View(available.packages())
```

As well as these repositories, there are many R packages in online repositories such as GitHub, Bitbucket, and Google Code. Retrieving packages from GitHub is particularly easy, as discussed below.

Many IDEs have a point-and-click method of installing packages. In R GUI, the Packages menu has the option "Install package(s)..." to install from a repository and "Install package(s) from local zip files..." to install packages that you downloaded earlier. Figure 10-3 shows the R GUI menu.

You can also install packages using the `install.packages` function. Calling it without any arguments gives you the same GUI interface as if you'd clicked the "Install package(s)..." menu option. Usually, you would want to specify the names of the packages that you want to download and the URL of the repository to retrieve them from. A list of URLs for CRAN mirrors is available on the main CRAN site (*http://cran.r-project.org/mirrors.html*).

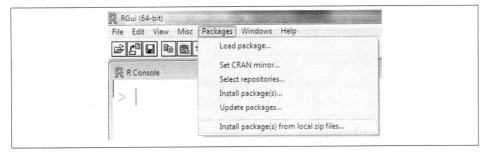

Figure 10-3. Installing packages in R GUI

This command will (try to) download the time-series analysis packages xts and zoo and all the dependencies for each, and then install them into the default library location (the first value returned by .libPaths):

```
install.packages(
  c("xts", "zoo"),
  repos = "http://www.stats.bris.ac.uk/R/"
)
```

To install to a different location, you can pass the lib argument to install.packages:

```
install.packages(
  c("xts", "zoo"),
  lib   = "some/other/folder/to/install/to",
  repos = "http://www.stats.bris.ac.uk/R/"
)
```

Obviously, you need a working Internet connection for R to be able to download packages, and you need sufficient permissions to be able to write files to the *library* folder. Inside corporate networks, R's access to the Internet may be restricted. Under Windows, you can get R to use *internet2.dll* to access the Internet, making it appear as though it is Internet Explorer and often bypassing restrictions. To achieve this, type:

```
setInternet2()
```

If all else fails, you can visit *http://<cran mirror>web/packages/available_packages_by_name.html* and manually download the packages that you want (remember to download all the dependencies too), then install the resultant *tar.gz/tgz/zip* file:

```
install.packages(
  "path/to/downloaded/file/xts_0.8-8.tar.gz",
  repos = NULL,      #NULL repo means "package already downloaded"
  type = "source"    #this means "build the package now"
)
```

```
install.packages(
  "path/to/downloaded/file/xts_0.8-8.zip",
  repos = NULL,          #still need this
  type = "win.binary" #Windows only!
)
```

To install a package directly from GitHub, you first need to install the `devtools` package:

```
install.packages("devtools")
```

The `install_github` function accepts the name of the GitHub repository that contains the package (usually the same as the name of the package itself) and the name of the user that maintains that repository. For example, to get the development version of the reporting package `knitr`, type:

```
library(devtools)
install_github("knitr", "yihui")
```

Maintaining Packages

After your packages are installed, you will usually want to update them in order to keep up with the latest versions. This is done with `update.packages`. By default, this function will prompt you before updating each package. This can become unwieldy after a while (having several hundred packages installed is not uncommon), so setting `ask = FALSE` is recommended:

```
update.packages(ask = FALSE)
```

Very occasionally, you may want to delete a package. It is possible to do this by simply deleting the folder containing the package contents from your filesystem, or you can do it programmatically:

```
remove.packages("zoo")
```

Summary

- There are thousands of R packages available from online repositories.

- You can install these packages with `install.packages`, and load them with `library` or `require`.

- When you load packages, they are added to the `search` path, which lets R find their variables.

- You can view the installed packages with `installed.packages`, keep them up-to-date with `update.packages`, and clean your system with `remove.packages`.

Test Your Knowledge: Quiz

Question 10-1
> What are the names of some R package repositories?

Question 10-2
> What is the difference between the `library` and `require` functions?

Question 10-3
> What is a package library?

Question 10-4
> How might you find the locations of package libraries on your machine?

Question 10-5
> How do you get R to pretend that it is Internet Explorer?

Test Your Knowledge: Exercises

Exercise 10-1
> Using R GUI, install the `Hmisc` package. [10]

Exercise 10-2
> Using the `install.packages` function, install the `lubridate` package. [10]

Exercise 10-3
> Count the number of packages that are installed on your machine in each library. [5]

## CHAPTER 11

Dates and Times

Dates and times are very common in data analysis—not least for time-series analysis. The bad news is that with different numbers of days in each month, leap years, leap seconds,[1] and time zones, they can be fairly awful to deal with programmatically. The good news is that R has a wide range of capabilities for dealing with times and dates. While these concepts are fairly fundamental to R programming, they've been left until now because some of the best ways of using them appear in add-on packages. As you begin reading this chapter, you may feel an awkward sensation that the code is grating on you. At this point, we'll seek lubrication from the lubridate package, which makes your date-time code more readable.

Chapter Goals

After reading this chapter, you should:

- Understand the built-in date classes POSIXct, POSIXlt, and Date
- Be able to convert a string into a date
- Know how to display dates in a variety of formats
- Be able to specify and manipulate time zones
- Be able to use the lubridate package

---

1. The spin of the Earth is slowing down, so it takes slightly longer than 86,400 seconds for a day to happen. This is especially obvious when you are waiting for payday. Leap seconds have been added since 1972 to correct for this. Type .leap.seconds to see when they have happened.

Date and Time Classes

There are three date and time classes that come with R: POSIXct, POSIXlt, and Date.

POSIX Dates and Times

POSIX dates and times are classic R: brilliantly thorough in their implementation, navigating all sorts of obscure technical issues, but with awful Unixy names that make everything seem more complicated than it really is.

The two standard date-time classes in R are POSIXct and POSIXlt. (I said the names were awful!) POSIX is a set of standards that defines compliance with Unix, including how dates and times should be specified. ct is short for "calendar time," and the POSIXct class stores dates as the number of seconds since the start of 1970, in the Coordinated Universal Time (UTC) zone.[2] POSIXlt stores dates as a list, with components for seconds, minutes, hours, day of month, etc. POSIXct is best for storing dates and calculating with them, whereas POSIXlt is best for extracting specific parts of a date.

The function Sys.time returns the current date and time in POSIXct form:

```
(now_ct <- Sys.time())
## [1] "2013-07-17 22:47:01 BST"
```

The class of now_ct has two elements. It is a POSIXct variable, and POSIXct is inherited from the class POSIXt:

```
class(now_ct)
## [1] "POSIXct" "POSIXt"
```

When a date is printed, you just see a formatted version of it, so it isn't obvious how the date is stored. By using unclass, we can see that it is indeed just a number:

```
unclass(now_ct)
## [1] 1.374e+09
```

When printed, the POSIXlt date looks exactly the same, but underneath the storage mechanism is very different:

```
(now_lt <- as.POSIXlt(now_ct))
## [1] "2013-07-17 22:47:01 BST"
class(now_lt)
## [1] "POSIXlt" "POSIXt"
```

2. UTC's acronym is the wrong way around to make it match other universal time standards (UT0, UT1, etc.). It is essentially identical to (civil) Greenwich Mean Time (GMT), except that Greenwich Mean Time isn't a scientific standard, and the British government can't change UTC.

```
unclass(now_lt)
## $sec
## [1] 1.19
##
## $min
## [1] 47
##
## $hour
## [1] 22
##
## $mday
## [1] 17
##
## $mon
## [1] 6
##
## $year
## [1] 113
##
## $wday
## [1] 3
##
## $yday
## [1] 197
##
## $isdst
## [1] 1
##
## attr(,"tzone")
## [1] ""      "GMT"  "BST"
```

You can use list indexing to access individual components of a POSIXlt date:

```
now_lt$sec
```

```
## [1] 1.19
```

```
now_lt[["min"]]
```

```
## [1] 47
```

The Date Class

The third date class in base R is slightly better-named: it is the Date class. This stores dates as the number of days since the start of 1970.[3] The Date class is best used when you don't care about the time of day. Fractional days are possible (and can be generated by calculating a mean Date, for example), but the POSIX classes are better for those situations:

3. Researchers of historical data might like to note that dates are always in the Gregorian calendar, so you need to double-check your code for anything before 1752.

```
(now_date <- as.Date(now_ct))
## [1] "2013-07-17"
class(now_date)
## [1] "Date"
unclass(now_date)
## [1] 15903
```

Other Date Classes

There are lots of other date and time classes scattered through other R classes. If you have a choice of which date-time class to use, you should usually stick to one of the three base classes (POSIXct, POSIXlt, and Date), but you need to be aware of the other classes if you are using other people's code that may depend upon them.

Other date and time classes from add-on packages include date, dates, chron, year mon, yearqtr, timeDate, ti, and jul.

Conversion to and from Strings

Many text file formats for data don't explicitly support specific date types. For example, in a CSV file, each value is just a string. In order to access date functionality in R, you must convert your date strings into variables of one of the date classes. Likewise, to write back to CSV, you must convert the dates back into strings.

Parsing Dates

When we read in dates from a text or spreadsheet file, they will typically be stored as a character vector or factor. To convert them to dates, we need to *parse* these strings. This can be done with another appallingly named function, strptime (short for "string parse time"), which returns POSIXlt dates. (There are as.POSIXct and as.POSIXlt functions too. If you call them on character inputs, then they are just wrappers around strptime.) To parse the dates, you must tell strptime which bits of the string correspond to which bits of the date. The date format is specified using a string, with components specified with a percent symbol followed by a letter. For example, the day of the month as a number is specified as %d. These components can be combined with other fixed characters—such as colons in times, or dashes and slashes in dates—to form a full specification. The time zone specification varies depending upon your operating system. It can get complicated, so the minutiae are discussed later, but you usually want "UTC" for universal time or "" to use the time zone in your current locale (as determined from your operating system's locale settings).

In the following example, %H is the hour (24-hour system), %M is the minute, %S is the second, %m is the number of the month, %d (as previously discussed) is the day of the month, and %Y is the four-digit year. The complete list of component specifiers varies from system to system. See the ?strptime help page for the details:

```
moon_landings_str <- c(
  "20:17:40 20/07/1969",
  "06:54:35 19/11/1969",
  "09:18:11 05/02/1971",
  "22:16:29 30/07/1971",
  "02:23:35 21/04/1972",
  "19:54:57 11/12/1972"
)
(moon_landings_lt <- strptime(
  moon_landings_str,
  "%H:%M:%S %d/%m/%Y",
  tz = "UTC"
))
## [1] "1969-07-20 20:17:40 UTC" "1969-11-19 06:54:35 UTC"
## [3] "1971-02-05 09:18:11 UTC" "1971-07-30 22:16:29 UTC"
## [5] "1972-04-21 02:23:35 UTC" "1972-12-11 19:54:57 UTC"
```

If a string does not match the format in the format string, it takes the value NA. For example, specifying dashes instead of slashes makes the parsing fail:

```
strptime(
  moon_landings_str,
  "%H:%M:%S %d-%m-%Y",
  tz = "UTC"
)
## [1] NA NA NA NA NA NA
```

Formatting Dates

The opposite problem of parsing is turning a date variable into a string—that is, formatting it. In this case, we use the same system for specifying a format string, but now we call strftime ("string format time") to reverse the parsing operation. In case you struggle to remember the name strftime, these days the format function will also happily format dates in a nearly identical manner to strftime.

In the following example, %I is the hour (12-hour system), %p is the AM/PM indicator, %A is the full name of the day of the week, and %B is the full name of the month. strftime works with both POSIXct and POSIXlt inputs:

```
strftime(now_ct, "It's %I:%M%p on %A %d %B, %Y.")
## [1] "It's 10:47PM on Wednesday 17 July, 2013."
```

Time Zones

Time zones are horrible, complicated things from a programming perspective. Countries often have several, and change the boundaries when some (but not all) switch to daylight savings time. Many time zones have abbreviated names, but they often aren't unique. For example, "EST" can refer to "Eastern Standard Time" in the United States, Canada, or Australia.

You can specify a time zone when parsing a date string (with `strptime`) and change it again when you format it (with `strftime`). During parsing, if you don't specify a time zone (the default is `""`), R will give the dates a default time zone. This is the value returned by `Sys.timezone`, which is in turn guessed from your operating system locale settings. You can see the OS date-time settings with `Sys.getlocale("LC_TIME")`.

The easiest way to avoid the time zone mess is to always record and then analyze your times in the UTC zone. If you can achieve this, congratulations! You are very lucky. For everyone else—those who deal with other people's data, for example—the easiest-to-read and most portable way of specifying time zones is to use the Olson form, which is "Continent/City" or similar:

```
strftime(now_ct, tz = "America/Los_Angeles")
## [1] "2013-07-17 14:47:01"
strftime(now_ct, tz = "Africa/Brazzaville")
## [1] "2013-07-17 22:47:01"
strftime(now_ct, tz = "Asia/Kolkata")
## [1] "2013-07-18 03:17:01"
strftime(now_ct, tz = "Australia/Adelaide")
## [1] "2013-07-18 07:17:01"
```

A list of possible Olson time zones is shipped with R in the file returned by `file.path(R.home("share"), "zoneinfo", "zone.tab")`. (That's a file called *zone.tab* in a folder called *zoneinfo* inside the *share* directory where you installed R.) The `lubridate` package described later in this chapter provides convenient access to this file.

The next most reliable method is to give a manual offset from UTC, in the form `"UTC"+n"` or `"UTC"-n"`. Negative times are east of UTC, and positive times are west. The manual nature at least makes it clear how the times are altered, but you have to manually do the daylight savings corrections too, so this method should be used with care. Recent versions of R will warn that the time zone is unknown, but will perform the offset correctly:

```
strftime(now_ct, tz = "UTC-5")
## Warning: unknown timezone 'UTC-5'
```

```
## [1] "2013-07-18 02:47:01"
strftime(now_ct, tz = "GMT-5")      #same
## Warning: unknown timezone 'GMT-5'
## [1] "2013-07-18 02:47:01"
strftime(now_ct, tz = "-5")         #same, if supported on your OS
## Warning: unknown timezone '-5'
## [1] "2013-07-18 02:47:01"
strftime(now_ct, tz = "UTC+2:30")
## Warning: unknown timezone 'UTC+2:30'
## [1] "2013-07-17 19:17:01"
```

The third method of specifying time zones is to use an abbreviation—either three letters or three letters, a number, and three more letters. This method is the last resort, for three reasons. First, abbreviations are harder to read, and thus more prone to errors. Second, as previously mentioned, they aren't unique, so you may not get the time zone that you think you have. Finally, different operating systems support different sets of abbreviations. In particular, the Windows OS's knowledge of time zone abbreviations is patchy:

```
strftime(now_ct, tz = "EST")      #Canadian Eastern Standard Time
## [1] "2013-07-17 16:47:01"
strftime(now_ct, tz = "PST8PDT")   #Pacific Standard Time w/ daylight savings
## [1] "2013-07-17 14:47:01"
```

One last word of warning about time zones: strftime ignores time zone changes for POSIXlt dates. It is best to explicitly convert your dates to POSIXct before printing:

```
strftime(now_ct, tz = "Asia/Tokyo")
## [1] "2013-07-18 06:47:01"
strftime(now_lt, tz = "Asia/Tokyo")              #no zone change!
## [1] "2013-07-17 22:47:01"
strftime(as.POSIXct(now_lt), tz = "Asia/Tokyo")
## [1] "2013-07-18 06:47:01"
```

Another last warning (really the last one!): if you call the concatenation function, c, with a POSIXlt argument, it will change the time zone to your local time zone. Calling c on a POSIXct argument, by contrast, will strip its time zone attribute completely. (Most other functions will assume that the date is now local, but be careful!)

Arithmetic with Dates and Times

R supports arithmetic with each of the three base classes. Adding a number to a POSIX date shifts it by that many seconds. Adding a number to a `Date` shifts it by that many days:

```
now_ct + 86400 #Tomorrow.  I wonder what the world will be like!

## [1] "2013-07-18 22:47:01 BST"

now_lt + 86400 #Same behavior for POSIXlt

## [1] "2013-07-18 22:47:01 BST"

now_date + 1   #Date arithmetic is in days

## [1] "2013-07-18"
```

Adding two dates together doesn't make much sense, and throws an error. Subtraction is supported, and calculates the difference between the two dates. The behavior is the same for all three date types. In the following example, note that `as.Date` will automatically parse dates of the form %Y-%m-%d or %Y/%m/%d, if you don't specify a format:

```
the_start_of_time <-     #according to POSIX
  as.Date("1970-01-01")
the_end_of_time <-       #according to Mayan conspiracy theorists
  as.Date("2012-12-21")
(all_time <- the_end_of_time - the_start_of_time)

## Time difference of 15695 days
```

We can use the now (hopefully) familiar combination of `class` and `unclass` to see how the difference in time is stored:

```
class(all_time)

## [1] "difftime"

unclass(all_time)

## [1] 15695
## attr(,"units")
## [1] "days"
```

The difference has class `difftime`, and the value is stored as a number with a unit attribute of days. Days were automatically chosen as the "most sensible" unit due to the difference between the times. Differences shorter than one day are given in hours, minutes, or seconds, as appropriate. For more control over the units, you can use the `difftime` function:

```
difftime(the_end_of_time, the_start_of_time, units = "secs")

## Time difference of 1.356e+09 secs
```

```
difftime(the_end_of_time, the_start_of_time, units = "weeks")

## Time difference of 2242 weeks
```

The seq function for generating sequences also works with dates. This can be particularly useful for creating test datasets of artificial dates. The choice of units in the by argument differs between the POSIX and Date types. See the ?seq.POSIXt and ?seq.Date help pages for the choices in each case:

```
seq(the_start_of_time, the_end_of_time, by = "1 year")

##  [1] "1970-01-01" "1971-01-01" "1972-01-01" "1973-01-01" "1974-01-01"
##  [6] "1975-01-01" "1976-01-01" "1977-01-01" "1978-01-01" "1979-01-01"
## [11] "1980-01-01" "1981-01-01" "1982-01-01" "1983-01-01" "1984-01-01"
## [16] "1985-01-01" "1986-01-01" "1987-01-01" "1988-01-01" "1989-01-01"
## [21] "1990-01-01" "1991-01-01" "1992-01-01" "1993-01-01" "1994-01-01"
## [26] "1995-01-01" "1996-01-01" "1997-01-01" "1998-01-01" "1999-01-01"
## [31] "2000-01-01" "2001-01-01" "2002-01-01" "2003-01-01" "2004-01-01"
## [36] "2005-01-01" "2006-01-01" "2007-01-01" "2008-01-01" "2009-01-01"
## [41] "2010-01-01" "2011-01-01" "2012-01-01"

seq(the_start_of_time, the_end_of_time, by = "500 days") #of Summer

##  [1] "1970-01-01" "1971-05-16" "1972-09-27" "1974-02-09" "1975-06-24"
##  [6] "1976-11-05" "1978-03-20" "1979-08-02" "1980-12-14" "1982-04-28"
## [11] "1983-09-10" "1985-01-22" "1986-06-06" "1987-10-19" "1989-03-02"
## [16] "1990-07-15" "1991-11-27" "1993-04-10" "1994-08-23" "1996-01-05"
## [21] "1997-05-19" "1998-10-01" "2000-02-13" "2001-06-27" "2002-11-09"
## [26] "2004-03-23" "2005-08-05" "2006-12-18" "2008-05-01" "2009-09-13"
## [31] "2011-01-26" "2012-06-09"
```

Many other base functions allow manipulation of dates. You can repeat them, round them, and cut them. You can also calculate summary statistics with mean and summary. Many of the possibilities can be seen with methods(class = "POSIXt") and methods(class = "Date"), although some other functions will handle dates without having specific date methods.

Lubridate

If you've become disheartened with dates and are considering skipping the rest of the chapter, do not fear! Help is at hand. lubridate, as the name suggests, adds some much-needed lubrication to the process of date manipulation. It doesn't add many new features over base R, but it makes your code more readable, and helps you avoid having to think too much.

To replace strptime, lubridate has a variety of parsing functions with predetermined formats. ymd accepts dates in the form year, month, day. There is some flexibility in the specification: several common separators like hyphens, forward and backward slashes, colons, and spaces can be used;[4] months can be specified by number or by full or abbreviated name; and the day of the week can optionally be included. The real beauty is that different elements in the same vector can have different formats (as long as the year is followed by the month, which is followed by the day):

```
library(lubridate)

## Attaching package: 'lubridate'

## The following object is masked from 'package:chron':
##
## days, hours, minutes, seconds, years

john_harrison_birth_date <- c( #He invented the marine chronometer
  "1693-03 24",
  "1693/03\\24",
  "Tuesday+1693.03*24"
)
ymd(john_harrison_birth_date)  #All the same

## [1] "1693-03-24 UTC" "1693-03-24 UTC" "1693-03-24 UTC"
```

The important thing to remember with ymd is to get the elements of the date in the right order. If your date data is in a different form, then lubridate provides other functions (ydm, mdy, myd, dmy, and dym) to use instead. Each of these functions has relatives that allow the specification of times as well, so you get ymd_h, ymd_hm, and ymd_hms, as well as the equivalents for the other five date orderings. If your dates aren't in any of these formats, then the lower-level parse_date_time lets you give a more exact specification.

All the parsing functions in lubridate return POSIXct dates and have a default time zone of UTC. Be warned: these behaviors are different from base R's strptime! (Although usually more convenient.) In lubridate terminology, these individual dates are "instants."

For formatting dates, lubridate provides stamp, which lets you specify a format in a more human-readable manner. You specify an example date, and it returns a function that you can call to format your dates:

```
date_format_function <-
  stamp("A moon landing occurred on Monday 01 January 1900 at 18:00:00.")

## Multiple formats matched: "A moon landing occurred on %A %m January %d%y
## at %H:%M:%OS"(1), "A moon landing occurred on %A %m January %Y at
## %d:%H:%M."(1), "A moon landing occurred on %A %d %B %Y at %H:%M:%S."(1)

## Using: "A moon landing occurred on %A %d %B %Y at %H:%M:%S."

date_format_function(moon_landings_lt)

## [1] "A moon landing occurred on Sunday 20 July 1969 at 20:17:40."
```

---

4. In fact, most punctuation is allowed.

For dealing with ranges of times, lubridate has three different variable types. "Durations" specify time spans as multiples of seconds, so a duration of a day is always 86,400 seconds (60 * 60 * 24), and a duration of a year is always 31,536,000 seconds (86,400 * 365). This makes it easy to specify ranges of dates that are exactly evenly spaced, but leap years and daylight savings time put them out of sync from clock time. In the following example, notice that the date slips back one day every time there is a leap year. today gives today's date:

```
(duration_one_to_ten_years <- dyears(1:10))
```

```
##  [1] "31536000s (~365 days)"   "63072000s (~730 days)"
##  [3] "94608000s (~1095 days)"  "126144000s (~1460 days)"
##  [5] "157680000s (~1825 days)" "189216000s (~2190 days)"
##  [7] "220752000s (~2555 days)" "252288000s (~2920 days)"
##  [9] "283824000s (~3285 days)" "315360000s (~3650 days)"
```

```
today() + duration_one_to_ten_years
```

```
##  [1] "2014-07-17" "2015-07-17" "2016-07-16" "2017-07-16" "2018-07-16"
##  [6] "2019-07-16" "2020-07-15" "2021-07-15" "2022-07-15" "2023-07-15"
```

Other functions for creating durations are dseconds, dminutes, and so forth, as well as new_duration for mixed-component specification.

"Periods" specify time spans according to clock time. That means that their exact length isn't apparent until you add them to an instant. For example, a period of one year can be 365 or 366 days, depending upon whether or not it is a leap year. In the following example, notice that the date stays the same across leap years:

```
(period_one_to_ten_years <- years(1:10))
```

```
##  [1] "1y 0m 0d 0H 0M 0S"  "2y 0m 0d 0H 0M 0S"  "3y 0m 0d 0H 0M 0S"
##  [4] "4y 0m 0d 0H 0M 0S"  "5y 0m 0d 0H 0M 0S"  "6y 0m 0d 0H 0M 0S"
##  [7] "7y 0m 0d 0H 0M 0S"  "8y 0m 0d 0H 0M 0S"  "9y 0m 0d 0H 0M 0S"
## [10] "10y 0m 0d 0H 0M 0S"
```

```
today() + period_one_to_ten_years
```

```
##  [1] "2014-07-17" "2015-07-17" "2016-07-17" "2017-07-17" "2018-07-17"
##  [6] "2019-07-17" "2020-07-17" "2021-07-17" "2022-07-17" "2023-07-17"
```

In addition to years, you can create periods with seconds, minutes, etc., as well as new_period for mixed-component specification.

"Intervals" are defined by the instants at their beginning and end. They aren't much use on their own—they are most commonly used for specifying durations and periods when you known the start and end dates (rather than how long they should last). They can also be used for converting between durations and periods. For example, given a duration of one year, direct conversion to a period can only be estimated, since periods of a year can be 365 or 366 days (possibly plus a few leap seconds, and possibly plus or minus an hour or two if the rules for daylight savings change):

```
a_year <- dyears(1)     #exactly 60*60*24*365 seconds
as.period(a_year)       #only an estimate
## estimate only: convert durations to intervals for accuracy
## [1] "1y 0m 0d 0H 0M 0S"
```

If we know the start (or end) date of the duration, we can use an `interval` and an intermediary to convert exactly from the duration to the period:

```
start_date <- ymd("2016-02-28")
(interval_over_leap_year <- new_interval(
  start_date,
  start_date + a_year
))
## [1] 2016-02-28 UTC--2017-02-27 UTC
as.period(interval_over_leap_year)
## [1] "11m 30d 0H 0M 0S"
```

Intervals also have some convenience operators, namely `%--%` for defining intervals and `%within%` for checking if a date is contained within an interval:

```
ymd("2016-02-28") %--% ymd("2016-03-01") #another way to specify interval
## [1] 2016-02-28 UTC--2016-03-01 UTC
ymd("2016-02-29") %within% interval_over_leap_year
## [1] TRUE
```

For dealing with time zones, `with_tz` lets you change the time zone of a date without having to print it (unlike `strftime`). It also correctly handles `POSIXlt` dates (again, unlike `strftime`):

```
with_tz(now_lt, tz = "America/Los_Angeles")
## [1] "2013-07-17 14:47:01 PDT"
with_tz(now_lt, tz = "Africa/Brazzaville")
## [1] "2013-07-17 22:47:01 WAT"
with_tz(now_lt, tz = "Asia/Kolkata")
## [1] "2013-07-18 03:17:01 IST"
with_tz(now_lt, tz = "Australia/Adelaide")
## [1] "2013-07-18 07:17:01 CST"
```

`force_tz` is a variant of `with_tz` used for updating incorrect time zones.

`olson_time_zones` returns a list of all the Olson-style time zone names that R knows about, either alphabetically or by longitude:

```
head(olson_time_zones())
## [1] "Africa/Abidjan"    "Africa/Accra"    "Africa/Addis_Ababa"
## [4] "Africa/Algiers"    "Africa/Asmara"   "Africa/Bamako"

head(olson_time_zones("longitude"))
## [1] "Pacific/Midway"    "America/Adak"       "Pacific/Chatham"
## [4] "Pacific/Wallis"    "Pacific/Tongatapu"  "Pacific/Enderbury"
```

Some other utilities are available for arithmetic with dates, particularly floor_date and ceiling_date:

```
floor_date(today(), "year")
## [1] "2013-01-01"

ceiling_date(today(), "year")
## [1] "2014-01-01"
```

Summary

- There are three built-in classes for storing dates and times: POSIXct, POSIXlt, and Date.
- Parsing turns a string into a date; it can be done with strptime.
- Formatting turns a date back into a string; it can be done with strftime.
- Time zones can be specified using an Olson name or an offset from UTC, or (sometimes) with a three-letter abbreviation.
- The lubridate package makes working with dates a bit easier.

Test Your Knowledge: Quiz

Question 11-1
Which of the three built-in classes would you use to store dates with times in a data frame?

Question 11-2
When is the origin (time 0) for POSIXct and Date dates?

Question 11-3
What formatting string would you use to display a date as the complete month name followed by the four-digit year number?

Question 11-4
How would you shift a POSIXct date one hour into the future?

Question 11-5

Using the `lubridate` package, consider two intervals starting on January 1, 2016. Which will finish later, a duration of one year or a period of one year?

Test Your Knowledge: Exercises

Exercise 11-1

Parse the birth dates of the Beatles, and print them in the form "AbbreviatedWeekday DayOfMonth AbbreviatedMonthName TwoDigitYear" (for example, "Wed 09 Oct 40"). Their dates of birth are given in the following table.

Beatle	Birth date
Ringo Starr	1940-07-07
John Lennon	1940-10-09
Paul McCartney	1942-06-18
George Harrison	1943-02-25

[10]

Exercise 11-2

Programmatically read the file that R uses to find Olson time zone names. The examples in the `?Sys.timezone` help page demonstrate how to do this. Find the name of the time zone for your location. [10]

Exercise 11-3

Write a function that accepts a date as an input and returns the astrological sign of the zodiac corresponding to that date. The date ranges for each sign are given in the following table. [15]

Zodiac sign	Start date	End date
Aries	March 21	April 19
Taurus	April 20	May 20
Gemini	May 21	June 20
Cancer	June 21	July 22
Leo	July 23	August 22
Virgo	August 23	September 22
Libra	September 23	October 22
Scorpio	October 23	November 21
Sagittarius	November 22	December 21
Capricorn	December 22	January 19
Aquarius	January 20	February 18
Pisces	February 19	March 20

## PART II

The Data Analysis Workflow

## CHAPTER 12

Getting Data

Data can come from many sources. R comes with many datasets built in, and there is more data in many of the add-on packages. R can read data from a wide variety of sources and in a wide variety of formats. This chapter covers importing data from text files (including spreadsheet-like data in comma- or tab-delimited format, XML, and JSON), binary files (Excel spreadsheets and data from other analysis software), websites, and databases.

Chapter Goals

After reading this chapter, you should:

- Be able to access datasets provided with R packages
- Be able to import data from text files
- Be able to import data from binary files
- Be able to download data from websites
- Be able to import data from a database

Built-in Datasets

One of the packages in the base R distribution is called `datasets`, and it is entirely filled with example datasets. While you'll be lucky if any of them are suited to your particular area of research, they are ideal for testing your code and for exploring new techniques. Many other packages also contain datasets. You can see all the datasets that are available in the packages that you have loaded using the `data` function:

```
data()
```

For a more complete list, including data from all packages that have been installed, use this invocation:

```
data(package = .packages(TRUE))
```

To access the data in any of these datasets, call the `data` function, this time passing the name of the dataset and the package where the data is found (if the package has been loaded, then you can omit the `package` argument):

```
data("kidney", package = "survival")
```

Now the `kidney` data frame can be used just like your own variables:

```
head(kidney)

##   id time status age sex disease frail
## 1  1    8      1  28   1   Other   2.3
## 2  1   16      1  28   1   Other   2.3
## 3  2   23      1  48   2      GN   1.9
## 4  2   13      0  48   2      GN   1.9
## 5  3   22      1  32   1   Other   1.2
## 6  3   28      1  32   1   Other   1.2
```

Reading Text Files

There are many, many formats and standards of text documents for storing data. Common formats for storing data are delimiter-separated values (CSV or tab-delimited), eXtensible Markup Language (XML), JavaScript Object Notation (JSON), and YAML (which recursively stands for YAML Ain't Markup Language). Other sources of text data are less well-structured—a book, for example, contains text data without any formal (that is, standardized and machine parsable) structure.

The main advantage of storing data in text files is that they can be read by more or less all other data analysis software and by humans. This makes your data more widely reusable by others.

CSV and Tab-Delimited Files

Rectangular (spreadsheet-like) data is commonly stored in delimited-value files, particularly comma-separated values (CSV) and tab-delimited values files. The `read.table` function reads these delimited files and stores the results in a data frame. In its simplest form, it just takes the path to a text file and imports the contents.

RedDeerEndocranialVolume.dlm is a whitespace-delimited file containing measurements of the endocranial volume of some red deer, measured using different techniques. (For those of you with an interest in deer skulls, the methods are computer tomography; filling the skull with glass beads; measuring the length, width, and height with calipers; and using Finarelli's equation. A second measurement was taken in some cases to get an idea of the accuracy of the techniques. I've been assured that the deer were already long dead before they had their skulls filled with beads!) The data file can be found inside the *extdata* folder in the `learningr` package. The first few rows are shown in Table 12-1.

Table 12-1. Sample data from RedDeerEndocranialVolume.dlm

SkullID	VolCT	VolBead	VolLWH	VolFinarelli	VolCT2	VolBead2	VolLWH2
DIC44	389	375	1484	337			
B11	389	370	1722	377			
DIC90	352	345	1495	328			
DIC83	388	370	1683	377			
DIC787	375	355	1458	328			
DIC1573	325	320	1363	291			
C120	346	335	1250	289	346	330	1264
C25	302	295	1011	250	303	295	1009
F7	379	360	1621	347	375	365	1647

The data has a header row, so we need to pass the argument `header = TRUE` to `read.table`. Since a second measurement wasn't always taken, not all the lines are complete. Passing the argument `fill = TRUE` makes `read.table` substitute NA values for the missing fields. The `system.file` function in the following example is used to locate files that are inside a package (in this case, the *RedDeerEndocranialVolume.dlm* file in the *extdata* folder of the package `learningr`).

```
library(learningr)
deer_file <- system.file(
  "extdata",
  "RedDeerEndocranialVolume.dlm",
  package = "learningr"
)
deer_data <- read.table(deer_file, header = TRUE, fill = TRUE)
str(deer_data, vec.len = 1)      #vec.len alters the amount of output

## 'data.frame':      33 obs. of  8 variables:
## $ SkullID     : Factor w/ 33 levels "A4","B11","B12",..: 14 ...
## $ VolCT       : int  389 389 ...
## $ VolBead     : int  375 370 ...
## $ VolLWH      : int  1484 1722 ...
## $ VolFinarelli: int  337 377 ...
## $ VolCT2      : int  NA NA ...
## $ VolBead2    : int  NA NA ...
## $ VolLWH2     : int  NA NA ...
```

```
head(deer_data)
```

```
##    SkullID VolCT VolBead VolLWH VolFinarelli VolCT2 VolBead2 VolLWH2
## 1   DIC44   389     375   1484          337     NA       NA      NA
## 2     B11   389     370   1722          377     NA       NA      NA
## 3   DIC90   352     345   1495          328     NA       NA      NA
## 4   DIC83   388     370   1683          377     NA       NA      NA
## 5  DIC787   375     355   1458          328     NA       NA      NA
## 6 DIC1573   325     320   1363          291     NA       NA      NA
```

Notice that the class of each column has been automatically determined, and row and column names have been automatically assigned. The column names are (by default) forced to be valid variable names (via make.names), and if row names aren't provided the rows are simply numbered 1, 2, 3, and so on.

There are lots of arguments to specify how the file will be read; perhaps the most important is sep, which determines the character to use as a separator between fields. You can also specify how many lines of data to read via nrow, and how many lines at the start of the file to skip. More advanced options include the ability to override the default row names, column names, and classes, and to specify the character encoding of the input file and how string input columns should be declared.

There are several convenience wrapper functions to read.table. read.csv sets the default separator to a comma, and assumes that the data has a header row. read.csv2 is its European cousin, using a comma for decimal places and a semicolon as a separator. Likewise read.delim and read.delim2 import tab-delimited files with full stops[1] or commas for decimal places, respectively.

Back in August 2008, scientists from the Centre for Environment, Fisheries, and Aquaculture Science (CEFAS) in Lowestoft, UK, attached a tag with a pressure sensor and a temperature sensor to a brown crab and dropped it into the North Sea. The crab then spent just over a year doing crabby things[2] before being caught by fishermen, who returned the tag to CEFAS.

The data from this tag is stored in a CSV file, along with some metadata. The first few rows of the file look like this:

---

1. Or periods, if you are American.

2. Specifically, migrating from the eastern North Sea near Germany to the western North Sea near the UK.

Comment :- clock reset to download data	
The following data are the ID block contents	
Firmware Version No	2
Firmware Build Level	70
The following data are the Tag notebook contents	
Mission Day	405
Last Deployment Date	08/08/2008 09:55:00
Deployed by Host Version	5.2.0
Downloaded by Host Version	6.0.0
Last Clock Set Date	05/01/2010 10:34:00
The following data are the Lifetime notebook contents	
Tag ID	A03401
Pressure Range	10
No of sensors	2

In this case, we can't just read everything with a single call to read.csv, since different pieces of data have different numbers of fields, and indeed *different fields*. We need to use the skip and nrow arguments of read.csv to specify which bits of the file to read:

```
crab_file <- system.file(
  "extdata",
  "crabtag.csv",
  package = "learningr"
)
(crab_id_block <- read.csv(
  crab_file,
  header = FALSE,
  skip = 3,
  nrow = 2
))
##                   V1 V2
## 1  Firmware Version No  2
## 2 Firmware Build Level 70

(crab_tag_notebook <- read.csv(
  crab_file,
  header = FALSE,
  skip = 8,
  nrow = 5
))
##                         V1                 V2
## 1               Mission Day                405
## 2       Last Deployment Date 08/08/2008 09:55:00
## 3   Deployed by Host Version              5.2.0
```

```
## 4 Downloaded by Host Version                    6.0.0
## 5          Last Clock Set Date 05/01/2010 10:34:00
(crab_lifetime_notebook <- read.csv(
  crab_file,
  header = FALSE,
  skip = 15,
  nrow = 3
))
##                  V1    V2
## 1           Tag ID A03401
## 2 Pressure Range     10
## 3  No of sensors      2
```

 The `colbycol` and `sqldf` packages contain functions that allow you to read part of a CSV file into R. This can provide a useful speed-up if you don't need all the columns or all the rows.

For really low-level control when importing this sort of file, you can use the `scan` function, on which `read.table` is based. Ordinarily, you should never have to resort to `scan`, but it can be useful for malformed or nonstandard files.

 If your data has been exported from another language, you may need to pass the `na.strings` argument to `read.table`. For data exported from SQL, use `na.strings = "NULL"`. For data exported from SAS or Stata, use `na.strings = "."`. For data exported from Excel, use `na.strings = c("", "#N/A", "#DIV/0!", "#NUM!")`.

The opposite task, writing files, is generally simpler than reading files, since you don't need to worry about oddities in the file—you usually want to create something standard. `read.table` and `read.csv` have the obviously named counterparts `write.table` and `write.csv`.

Both functions take a data frame and a file path to write to. They also provide a few options to customize the output (whether or not to include row names and what the character encoding of the output file should be, for example):

```
write.csv(
  crab_lifetime_notebook,
  "Data/Cleaned/crab lifetime data.csv",
  row.names    = FALSE,
  fileEncoding = "utf8"
)
```

Unstructured Text Files

Not all text files have a well-defined structure like delimited files do. If the structure of the file is weak, it is often easier to read in the file as lines of text and then parse or manipulate the contents afterward. readLines (notice the capital "L") provides such a facility. It accepts a path to a file (or a file connection) and, optionally, a maximum number of lines to read. Here we import the Project Gutenberg version of Shakespeare's *The Tempest*:

```
text_file <- system.file(
  "extdata",
  "Shakespeare's The Tempest, from Project Gutenberg pg2235.txt",
  package = "learningr"
)
the_tempest <- readLines(text_file)
the_tempest[1926:1927]

## [1] "   Ste. Foure legges and two voyces; a most delicate"
## [2] "Monster: his forward voyce now is to speake well of"
```

writeLines performs the reverse operation to readLines. It takes a character vector and a file to write to:

```
writeLines(
  rev(text_file),       #rev reverses vectors
  "Shakespeare's The Tempest, backwards.txt"
)
```

XML and HTML Files

XML files are widely used for storing nested data. Many standard file types and protocols are based upon it, such as RSS (Really Simple Syndication) for news feeds, SOAP (Simple Object Access Protocol) for passing structured data across computer networks, and the XHTML flavor of web pages.

Base R doesn't ship with the capability to read XML files, but the XML package is developed by an R Core member. Install it now!

```
install.packages("XML")
```

When you import an XML file, the XML package gives you a choice of storing the result using internal nodes (that is, objects are stored with C code, the default) or R nodes. Usually you want to store things with internal nodes, because it allows you to query the node tree using XPath (more on this in a moment).

There are several functions for importing XML data, such as xmlParse and some other wrapper functions that use slightly different defaults:

```
library(XML)
xml_file <- system.file("extdata", "options.xml", package = "learningr")
r_options <- xmlParse(xml_file)
```

One of the problems with using internal nodes is that summary functions like str and head don't work with them. To use R-level nodes, set useInternalNodes = FALSE (or use xmlTreeParse, which sets this as the default):

```
xmlParse(xml_file, useInternalNodes = FALSE)
xmlTreeParse(xml_file)        #the same
```

XPath is a language for interrogating XML documents, letting you find nodes that correspond to some filter. In this next example we look anywhere in the document (//) for a node named variable where ([ ]) the name attribute (@) contains the string *warn*.

```
xpathSApply(r_options, "//variable[contains(@name, 'warn')]")
## [[1]]
## <variable name="nwarnings" type="numeric">
##    <value>50</value>
## </variable>
##
## [[2]]
## <variable name="warn" type="numeric">
##    <value>0</value>
## </variable>
##
## [[3]]
## <variable name="warning_length" type="numeric">
##    <value>1000</value>
## </variable>
```

This sort of querying is very useful for extracting data from web pages. The equivalent functions for importing HTML pages are named as you might expect, htmlParse and htmlTreeParse, and they behave in the same way.

XML is also very useful for serializing (a.k.a. "saving") objects in a format that can be read by most other pieces of software. The XML package doesn't provide serialization functionality, but it is available via the makexml function in the Runiversal package. The *options.xml* file was created with this code:

```
library(Runiversal)
ops <- as.list(options())
cat(makexml(ops), file = "options.xml")
```

JSON and YAML Files

The main problems with XML are that it is very verbose, and you need to explicitly specify the type of the data (it can't tell the difference between a string and a number by default), which makes it even more verbose. When file sizes are important (such as when you are transferring big datasets across a network), this verbosity becomes problematic.

YAML and its subset JSON were invented to solve these problems. They are much better suited to transporting many datasets—particularly numeric data and arrays—over networks. JSON is the de facto standard for web applications to pass data between themselves.

There are two packages for dealing with JSON data: RJSONIO and rjson. For a long time rjson had performance problems, so the only package that could be recommended was RJSONIO. The performance issues have now been fixed, so it's a much closer call. For most cases, it now doesn't matter which package you use. The differences occur when you encounter malformed or nonstandard JSON.

RJSONIO is generally more forgiving than rjson when reading incorrect JSON. Whether this is a good thing or not depends upon your use case. If you think it is better to import JSON data with minimal fuss, then RJSONIO is best. If you want to be alerted to problems with the JSON data (perhaps it was generated by a colleague—I'm sure you would never generate malformed JSON), then rjson is best.

Fortunately, both packages have identically named functions for reading and writing JSON data, so it is easy to swap between them. In the following example, the double colons, ::, are used to distinguish which package each function should be taken from (if you only load one of the two packages, then you don't need the double colons):

```
library(RJSONIO)
library(rjson)
jamaican_city_file <- system.file(
  "extdata",
  "Jamaican Cities.json",
  package = "learningr"
)
(jamaican_cities_RJSONIO <- RJSONIO::fromJSON(jamaican_city_file))

## $Kingston
## $Kingston$population
## [1] 587798
##
## $Kingston$coordinates
## longitude  latitude
##     17.98     76.80
##
##
## $`Montego Bay`
## $`Montego Bay`$population
## [1] 96488
##
## $`Montego Bay`$coordinates
## longitude  latitude
##     18.47     77.92
```

```
(jamaican_cities_rjson <- rjson::fromJSON(file = jamaican_city_file))
## $Kingston
## $Kingston$population
## [1] 587798
##
## $Kingston$coordinates
## $Kingston$coordinates$longitude
## [1] 17.98
##
## $Kingston$coordinates$latitude
## [1] 76.8
##
##
##
## $`Montego Bay`
## $`Montego Bay`$population
## [1] 96488
##
## $`Montego Bay`$coordinates
## $`Montego Bay`$coordinates$longitude
## [1] 18.47
##
## $`Montego Bay`$coordinates$latitude
## [1] 77.92
```

Notice that RJSONIO simplifies the coordinates for each city to be a vector. This behavior can be turned off with `simplify = FALSE`, resulting in exactly the same object as the one generated by rjson.

Annoyingly, the JSON spec doesn't allow infinite or NaN values, and it's a little fuzzy on what a missing number should look like. The two packages deal with these values differently—RJSONIO maps NaN and NA to JSON's null value but preserves positive and negative infinity, while rjson converts all these values to strings:

```
special_numbers <- c(NaN, NA, Inf, -Inf)
RJSONIO::toJSON(special_numbers)

## [1] "[ null, null,    Inf,   -Inf ]"

rjson::toJSON(special_numbers)

## [1] "[\"NaN\",\"NA\",\"Inf\",\"-Inf\"]"
```

Since both these methods are hacks to deal with JSON's limited spec, if you find yourself dealing with these special number types a lot (or want to write comments in your data object), then you are better off using YAML. The yaml package has two functions for importing YAML data. yaml.load accepts a string of YAML and converts it to an R object, and yaml.load_file does the same, but treats its string input as a path to a file containing YAML:

```
library(yaml)
yaml.load_file(jamaican_city_file)
```

```
## $Kingston
## $Kingston$population
## [1] 587798
##
## $Kingston$coordinates
## $Kingston$coordinates$longitude
## [1] 17.98
##
## $Kingston$coordinates$latitude
## [1] 76.8
##
##
##
## $`Montego Bay`
## $`Montego Bay`$population
## [1] 96488
##
## $`Montego Bay`$coordinates
## $`Montego Bay`$coordinates$longitude
## [1] 18.47
##
## $`Montego Bay`$coordinates$latitude
## [1] 77.92
```

`as.yaml` performs the opposite task, converting R objects to YAML strings.

Reading Binary Files

Many pieces of software store their data in binary formats (some proprietary, some conforming to publically defined standards). The binary formats are often smaller than their text equivalents, so performance gains are usually possible by using a binary format, at the expense of human readability.

Many binary file formats are proprietary, which goes against free software principles. If you have the option, it is usually best to avoid such formats in order to stop your data being locked into a platform over which you lack control.

Reading Excel Files

Microsoft Excel is the world's most popular spreadsheet package, and very possibly the world's most popular data analysis tool. Unfortunately, its document formats, XLS and XLSX, don't play very nicely with other software, especially outside of a Windows environment. This means that some experimenting may be required to find a setup that works for your choice of operating system and the particular type of Excel file.

That said, the xlsx package is Java-based and cross-platform, so at least in theory it can read any Excel file on any system. It provides a choice of functions for reading Excel files: spreadsheets can be imported with read.xlsx and read.xlsx2, which do more processing in R and in Java, respectively. The choice of two engines is rather superfluous; you want read.xlsx2, since it's faster and the underlying Java library is more mature than the R code.

The next example features the fastest times from the Alpe d'Huez mountain stage of the Tour de France bike race, along with whether or not each cyclist has been found guilty of using banned performance-enhancing drugs. The colClasses argument determines what class each column should have in the resulting data frame. It isn't compulsory, but it saves you having to manipulate the resulting data afterward:

```
library(xlsx)
bike_file <- system.file(
  "extdata",
  "Alpe d'Huez.xls",
  package = "learningr"
)
bike_data <- read.xlsx2(
  bike_file,
  sheetIndex = 1,
  startRow   = 2,
  endRow     = 38,
  colIndex   = 2:8,
  colClasses = c(
    "character", "numeric", "character", "integer",
    "character", "character", "character"
  )
)
head(bike_data)
```

```
##       Time NumericTime            Name Year    Nationality DrugUse
## 1 37' 35"       37.58  Marco Pantani 1997          Italy       Y
## 2 37' 36"       37.60 Lance Armstrong 2004 United States       Y
## 3 38' 00"       38.00  Marco Pantani 1994          Italy       Y
## 4 38' 01"       38.02 Lance Armstrong 2001 United States       Y
## 5 38' 04"       38.07  Marco Pantani 1995          Italy       Y
## 6 38' 23"       38.38     Jan Ullrich 1997        Germany       Y
##                                                      Allegations
## 1 Alleged drug use during 1997 due to high haematocrit levels.
## 2         2004 Tour de France title stripped by USADA in 2012.
## 3 Alleged drug use during 1994 due to high haematocrit levels.
## 4         2001 Tour de France title stripped by USADA in 2012.
## 5 Alleged drug use during 1995 due to high haematocrit levels.
## 6                 Found guilty of a doping offense by the CAS.
```

The xlsReadWrite package provides an alternative to xlsx, but it currently only works with 32-bit R installations, and only on Windows. There are also some other packages designed to work with Excel. RExcel and excel.link use COM connections to control Excel from R, and WriteXLS uses Perl to write to Excel files. The gnumeric package provides functions for reading Gnumeric spreadsheets.

The counterpart to read.xlsx2 is (you guessed it) write.xlsx2. It works the same way as write.csv, taking a data frame and a filename. Unless you really need to use Excel spreadsheets, you are better off saving your data in a text format for portability, so use this with caution.

Reading SAS, Stata, SPSS, and MATLAB Files

If you are collaborating with statisticians in other organizations, they may try to send you files from another statistical package. The foreign package contains methods to read SAS permanent datasets[3] (SAS7BDAT files) using read.ssd, Stata DTA files with read.dta, and SPSS data files with read.spss. Each of these files can be written with write.foreign.

MATLAB binary data files (Level 4 and Level 5) can be read and written using read Mat and writeMat in the R.matlab package.

Reading Other File Types

R can read data from many other types of files.

It can read Hierarchical Data Format v5 [HDF5 (*http://www.hdfgroup.org/HDF5/*)] files via the h5r package (and the rdhf5 package on Bioconductor), and Network Common Data Format [NetCDF (*http://www.unidata.ucar.edu/software/netcdf/*)] files via the ncdf package.

It can read ESRI ArcGIS spatial data files via the maptools and shapefiles packages (and older ArcInfo files via the RArcInfo package).

It can read raster picture formats via the jpeg, png, tiff, rtiff, and readbitmap packages.

It can read a variety of genomics data formats using packages on Bioconductor. Most notably, it can read microarray data in GenePix GPR files (a.k.a. axon text files) via the RPPanalyzer package; Variant Call Format (VCF) files for gene sequence variants via

---

3. There isn't currently a method of importing the older SAS SD2 files into R. The easiest way of dealing with a file in this format is to open it with the free SAS Universal Viewer (*http://bit.ly/1dqSmrB*) and resave it as CSV.

the vcf2geno package; binary sequence alignment data via rbamtools (which provides an interface to SAMtools); and Luminex bead array assay files via the lxb package.

Finally, there are loads of miscellaneous formats scattered in other packages. A nonexhaustive list includes the 4dfp and tractor.base packages for MRI images, the IgorR package for WaveMetrics Igor binary format files, the GENEAread package for GENEActiv watch accelerometer data, the emme2 package for Inro Software EMME v2 databank files, the SEER2R package for SEER cancer datasets (*http://seer.cancer.gov/seerstat*), the rprotobuf package for Google's protocol buffers (*http://code.google.com/p/protobuf*), the readBrukerFlexData package for mass spectrometry data in Bruker flex format (*http://strimmerlab.org/software/maldiquant/*), the M3 package for community multiscale air quality models in Models-3 files, and the Read.isi package for ISI codebooks from the World Fertility Survey (*http://bit.ly/15xrAIa*).

While most of these packages are completely useless to most people, the fact that R can access so many niches in so many different areas is rather astounding.

Web Data

The Internet contains a wealth of data, but it's hard work (and not very scalable) to manually visit a website, download a data file, and then read it into R from your hard drive.

Fortunately, R has a variety of ways to import data from web sources; retrieving the data programmatically makes it possible to collect much more of it with much less effort.

Sites with an API

Several packages exist that download data directly into R using a website's application programming interface (API). For example, the World Bank makes its World Development Indicators data (*http://data.worldbank.org/data-catalog/world-development-indicators*) publically available, and the WDI package lets you easily import the data without leaving R. To run the next example, you first need to install the WDI package:

```
install.packages("WDI")

library(WDI)
#list all available datasets
wdi_datasets <- WDIsearch()
head(wdi_datasets)

##      indicator
## [1,] "BG.GSR.NFSV.GD.ZS"
## [2,] "BM.KLT.DINV.GD.ZS"
## [3,] "BN.CAB.XOKA.GD.ZS"
## [4,] "BN.CUR.GDPM.ZS"
## [5,] "BN.GSR.FCTY.CD.ZS"
```

```
## [6,] "BN.KLT.DINV.CD.ZS"
##       name
## [1,] "Trade in services (% of GDP)"
## [2,] "Foreign direct investment, net outflows (% of GDP)"
## [3,] "Current account balance (% of GDP)"
## [4,] "Current account balance excluding net official capital grants (% of GDP)"
## [5,] "Net income (% of GDP)"
## [6,] "Foreign direct investment (% of GDP)"

#retrieve one of them
wdi_trade_in_services <- WDI(
  indicator = "BG.GSR.NFSV.GD.ZS"
)

str(wdi_trade_in_services)

## 'data.frame':  984 obs. of  4 variables:
##  $ iso2c            : chr  "1A" "1A" "1A" ...
##  $ country          : chr  "Arab World" "Arab World" "Arab World" ...
##  $ BG.GSR.NFSV.GD.ZS: num  17.5 NA NA NA ...
##  $ year             : num  2005 2004 2003 2002 ...
```

The SmarterPoland package provides a similar wrapper to Polish government data. quantmod provides access to stock tickers (Yahoo!'s by default, though several other sources are available):

```
library(quantmod)
#If you are using a version before 0.5.0 then set this option
#or pass auto.assign = FALSE to getSymbols.
options(getSymbols.auto.assign = FALSE)
microsoft <- getSymbols("MSFT")

head(microsoft)

##            MSFT.Open MSFT.High MSFT.Low MSFT.Close MSFT.Volume
## 2007-01-03     29.91     30.25    29.40      29.86    76935100
## 2007-01-04     29.70     29.97    29.44      29.81    45774500
## 2007-01-05     29.63     29.75    29.45      29.64    44607200
## 2007-01-08     29.65     30.10    29.53      29.93    50220200
## 2007-01-09     30.00     30.18    29.73      29.96    44636600
## 2007-01-10     29.80     29.89    29.43      29.66    55017400
##            MSFT.Adjusted
## 2007-01-03         25.83
## 2007-01-04         25.79
## 2007-01-05         25.64
## 2007-01-08         25.89
## 2007-01-09         25.92
## 2007-01-10         25.66
```

The twitteR package provides access to Twitter's users and their tweets. There's a little bit of setup involved (due to Twitter's API requiring you to create an application and register it using OAuth; read the vignette for the package for setup instructions), but after that the package makes it easy to import Twitter data for network analysis, or simply look at tweets while pretending to work.

Scraping Web Pages

R has its own web server built in, so some functions for reading data are Internet-enabled by default. `read.table` (and its derivatives, like `read.csv`) can accept a URL rather than a local file, and will download a copy to a temp file before importing the data. For example, economic researcher Justin Rao's website has NBA basketball salary data for 2002 through 2008:

```
salary_url <- "http://www.justinmrao.com/salary_data.csv"
salary_data <- read.csv(salary_url)
str(salary_data)
```

Since accessing a large file over the Internet can be slow, if the file is going to be used often, a better strategy is to download the file using `download.file` to create a local copy, and then import that:

```
salary_url <- "http://www.justinmrao.com/salary_data.csv"
local_copy <- "my local copy.csv"
download.file(salary_url, local_copy)
salary_data <- read.csv(local_copy)
```

More advanced access to web pages can be achieved through the `RCurl` package, which provides access to the `libcurl` network client interface library. This is particularly useful if your data is contained inside an HTML or XML page, rather than a standard data format (like CSV) that just happens to be on the Web.

The next example retrieves the current date and time in several time zones from the United States Naval Observatory Time Service Department web page. The function `getURL` retrieves the page as a character string:

```
library(RCurl)
time_url <- "http://tycho.usno.navy.mil/cgi-bin/timer.pl"
time_page <- getURL(time_url)
cat(time_page)

## <!DOCTYPE HTML PUBLIC "-//W3C//DTD HTML 3.2 Final"//EN>
## <html>
## <body>
## <TITLE>What time is it?</TITLE>
## <H2> US Naval Observatory Master Clock Time</H2> <H3><PRE>
## <BR>Jul. 17, 20:43:37 UTC              Universal Time
## <BR>Jul. 17, 04:43:37 PM EDT Eastern Time
## <BR>Jul. 17, 03:43:37 PM CDT Central Time
## <BR>Jul. 17, 02:43:37 PM MDT Mountain Time
## <BR>Jul. 17, 01:43:37 PM PDT Pacific Time
## <BR>Jul. 17, 12:43:37 PM AKDT         Alaska Time
## <BR>Jul. 17, 10:43:37 AM HAST         Hawaii-Aleutian Time
## </PRE></H3><P><A HREF="http://www.usno.navy.mil"> US Naval Observatory</A>
##
## </body></html>
```

The next step is almost always to parse the page using htmlParse (or a related function) from the XML package. That allows you to extract the useful nodes. In the next example, splitting on \n (newlines) retrieves each time line, and splitting on \t (tabs) retrieves time/time zone pairs:

```
library(XML)
time_doc <- htmlParse(time_page)
pre <- xpathSApply(time_doc, "//pre")[[1]]
values <- strsplit(xmlValue(pre), "\n")[[1]][-1]
strsplit(values, "\t+")
## [[1]]
## [1] "Jul. 17, 20:43:37 UTC" "Universal Time"
##
## [[2]]
## [1] "Jul. 17, 04:43:37 PM EDT" "Eastern Time"
##
## [[3]]
## [1] "Jul. 17, 03:43:37 PM CDT" "Central Time"
##
## [[4]]
## [1] "Jul. 17, 02:43:37 PM MDT" "Mountain Time"
##
## [[5]]
## [1] "Jul. 17, 01:43:37 PM PDT" "Pacific Time"
##
## [[6]]
## [1] "Jul. 17, 12:43:37 PM AKDT" "Alaska Time"
##
## [[7]]
## [1] "Jul. 17, 10:43:37 AM HAST" "Hawaii-Aleutian Time"
```

The httr package is based on RCurl and provides syntactic sugar to make some tasks go down easier. The httr equivalent of RCurl's getURL is GET, and the content function retrieves the page's content, parsing it in the process. In the next example, we pass useInternalNodes = TRUE to mimic the behavior of htmlParse and repeat the action of the previous example:

```
library(httr)
time_page <- GET(time_url)
time_doc <- content(page, useInternalNodes = TRUE)
```

Accessing Databases

Where data has to be accessed by many people, it is often best stored in a relational database. There are many database management systems (DBMSs) for working with relational databases, and R can connect to all the common ones. The DBI package provides a unified syntax for accessing several DBMSs—currently SQLite, MySQL/MariaDB, PostgreSQL, and Oracle are supported, as well as a wrapper to the Java

Database Connectivity (JDBC) API. (Connections to SQL Server use a different system, as we'll see below.)

To connect to an SQLite database, you must first install and load the DBI package and the backend package RSQLite:

```
library(DBI)
library(RSQLite)
```

Then you define the database driver to be of type "SQLite" and set up a connection to the database, in this case by naming the file:

```
driver <- dbDriver("SQLite")
db_file <- system.file(
  "extdata",
  "crabtag.sqlite",
  package = "learningr"
)
conn <- dbConnect(driver, db_file)
```

The equivalent for a MySQL database would be to load the RMySQL package and set the driver type to be "MySQL":

```
driver <- dbDriver("MySQL")
db_file <- "path/to/MySQL/database"
conn <- dbConnect(driver, db_file)
```

The PostgreSQL, Oracle, and JDBC equivalents require the PostgreSQL, ROracle, and RJDBC packages, respectively, and their database names are the driver names, just as with SQLite and MySQL.

To retrieve data from the databases you write a query, which is just a string containing SQL commands, and send it to the database with dbGetQuery. In this next example, SELECT * FROM IdBlock means "get every column of the table named IdBlock":[4]

```
query <- "SELECT * FROM IdBlock"
(id_block <- dbGetQuery(conn, query))

##   Tag ID Firmware Version No Firmware Build Level
## 1 A03401                  2                    70
```

Then, after you've finished manipulating the database, you need to clean up by disconnecting and unloading the driver:

```
dbDisconnect(conn)

## [1] TRUE
```

---

4. Any database administrators worth their salt will tell you that writing SELECT * is lazy, and you should jolly well explicitly name your columns before that code goes anywhere near their servers, thank you very much.

```
dbUnloadDriver(driver)

## [1] TRUE
```

It is very easy to accidentally leave connections open, especially if an error occurs while you are connected. One way to avoid this is to wrap your database code into a function, and use on.exit to make sure that the cleanup code always runs. on.exit runs R code whenever its parent function exits, whether it finishes correctly or throws an error. We can rewrite the previous example with safer code as follows:

```
query_crab_tag_db <- function(query)
{
  driver <- dbDriver("SQLite")
  db_file <- system.file(
    "extdata",
    "crabtag.sqlite",
    package = "learningr"
  )
  conn <- dbConnect(driver, db_file)
  on.exit(
    {
      #this code block runs at the end of the function,
      #even if an error is thrown
      dbDisconnect(conn)
      dbUnloadDriver(driver)
    }
  )
  dbGetQuery(conn, query)
}
```

We can pass any SQL code to the function to query the crab tag database:

```
query_crab_tag_db("SELECT * FROM IdBlock")

##   Tag ID Firmware Version No Firmware Build Level
## 1 A03401                2                     70
```

In this case, the DBI package provides a convenience function that saves us having to write our own SQL code. dbReadTable does as you might expect: it reads a table from the connected database (use dbListTables(conn) if you can't remember the name of the table that you want):

```
dbReadTable(conn, "idblock")

##   Tag.ID Firmware.Version.No Firmware.Build.Level
## 1 A03401                   2                   70

## [1] TRUE

## [1] TRUE
```

If your database isn't one of the types listed here, the RODBC package is an alternative that uses ODBC database connections—this is particularly useful for connecting to SQL Server or Access databases. The functions have different names than their DBI counterparts, but the principles are very similar. You need to set up an ODBC data source on your machine (via the Control Panel under Windows; search for "ODBC" in the Start Menu) before R can connect. Figure 12-1 shows the OBDC registration wizard from Windows 7.

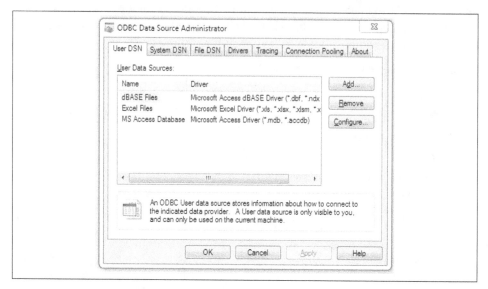

Figure 12-1. Registering an ODBC data source with the ODBC Data Source Administrator

Then call odbcConnect to connect, sqlQuery to run a query, and odbcClose to clean up afterward:

```
library(RODBC)
conn <- odbcConnect("my data source name")
id_block <- sqlQuery(conn, "SELECT * FROM IdBlock")
odbcClose(conn)
```

Methods for accessing NoSQL databases (short for "Not only SQL"; lightweight data stores that are more easily scalable than traditional SQL relational databases) from R are less mature. MongoDB can be accessed via the RMongo or rmongodb packages, Cassandra can be accessed from the RCassandra package, and CouchDB can be accessed from the R4CouchDB package [not yet on CRAN but available on GitHub (*https://github.com/wactbprot/R4CouchDB*)].

Summary

- Datasets supplied with R or packages can be made available with the data function.

- You can import data into R from a very wide range of external sources.
- `read.table` and its variants read rectangular data.
- `readLines` reads text files with nonstandard structures.
- You can read HTML and XML data with the `XML` package.
- The `RJSONIO`, `rjson`, and `yaml` packages read JSON/YAML.
- There are lots of packages for reading Excel files, including `xlsx`.
- The `foreign` package reads data from other statistics software.
- There are lots of packages for manipulating databases, including `DBI` and `RODBC`.

Test Your Knowledge: Quiz

Question 12-1
> How do you find all the datasets built into R and the packages on your machine?

Question 12-2
> What is the difference between the `read.csv` and `read.csv2` functions?

Question 12-3
> How would you import data from an Excel spreadsheet into R?

Question 12-4
> How would you import data from a CSV file found on the Internet?

Question 12-5
> The `DBI` package provides a consistent interface to several database management systems. Which ones are supported?

Test Your Knowledge: Exercises

Exercise 12-1
> In the *extdata* folder of the `learningr` package, there is a file named *hafu.csv*, containing data on mixed-race manga characters. Import the data into a data frame.
> [5]

Exercise 12-2
> Also in the *extdata* folder of `learningr` is an Excel file named *multi-drug-resistant gonorrhoea infection.xls*. Import the data from the first (and only) sheet into a data frame. Hint: this is a little easier if you view the file in a spreadsheet program first. LibreOffice (*https://www.libreoffice.org*) is free and accomplishes this task easily.
> [10]

Exercise 12-3

From the crab tag SQLite database described in this chapter, import the contents of the DayLog table into a data frame. [10]

## CHAPTER 13

Cleaning and Transforming

No matter what format you are given data in, it's almost always the wrong one for what you want to do with it, and no matter who gave it to you, it's almost always dirty. Cleaning and transforming data may not be the fun part of data analysis, but you'll probably spend more of your life than you care to doing it. Fortunately, R has a wide selection of tools to help with these tasks.

Chapter Goals

After reading this chapter, you should:

- Know how to manipulate strings and clean categorical variables
- Be able to subset and transform data frames
- Be able to change the shape of a data frame from wide to long and back again
- Understand sorting and ordering

Cleaning Strings

Back in Chapter 7, we looked at some simple string manipulation tasks like combining strings together using `paste`, and extracting sections of a string using `substring`.

One really common problem is when logical values have been encoded in a way that R doesn't understand. In the `alpe_d_huez` cycling dataset, the `DrugUse` column (denoting whether or not allegations of drug use have been made about each rider's performance), values have been encoded as `"Y"` and `"N"` rather than `TRUE` or `FALSE`. For this sort of simple matching, we can directly replace each string with the correct logical value:

```
yn_to_logical <- function(x)
{
  y <- rep.int(NA, length(x))
  y[x == "Y"] <- TRUE
  y[x == "N"] <- FALSE
  y
}
```

Setting values to NA by default lets us deal with strings that don't match "Y" or "N". We call the function in the obvious way:

```
alpe_d_huez$DrugUse <- yn_to_logical(alpe_d_huez$DrugUse)
```

This direct replacement of one string with another doesn't scale very well to having lots of choices of string. If you have ten thousand possible inputs, then a function to change each one would be very hard to write without errors, and even harder to maintain.

Fortunately, much more sophisticated manipulation is possible, and it is relatively easy to detect, extract, and replace parts of strings that match a pattern. R has a suite of built-in functions for handling these tasks, (loosely) based upon the Unix grep tool. They accept a string to manipulate and a regular expression to match. As mentioned in Chapter 1, regular expressions are patterns that provide a flexible means of describing the contents of a string. They are very useful for matching complex string-data types like phone numbers or email addresses.[1]

The grep, grepl, and regexpr functions all find strings that match a pattern, and sub and gsub replace matching strings. In classic R style, these functions are meticulously correct and very powerful, but suffer from funny naming conventions, quirky argument ordering, and odd return values that have arisen for historical reasons. Fortunately, in the same way that plyr provides a consistent wrapper around apply functions and lubridate provides a consistent wrapper around the date-time functions, the stringr package provides a consistent wrapper around the string manipulation functions. The difference is that while you will occasionally need to use a base apply or date-time function, stringr is advanced enough that you shouldn't need to bother with grep at all. So, take a look at the ?grep help page, but don't devote too much of your brain to it.

These next examples use the english_monarchs dataset from the learningr package. It contains the names and dates of rulers from post-Roman times (in the fifth century CE), when England was split into seven regions known as the heptarchy, until England took over Ireland in the early thirteenth century:

```
data(english_monarchs, package = "learningr")
head(english_monarchs)
```

---

1. See the assertive package for some pre-canned regular expressions for this purpose.

```
##       name       house start.of.reign end.of.reign       domain
## 1    Wehha  Wuffingas             NA          571 East Anglia
## 2    Wuffa  Wuffingas            571          578 East Anglia
## 3   Tytila  Wuffingas            578          616 East Anglia
## 4  Rædwald  Wuffingas            616          627 East Anglia
## 5 Eorpwald  Wuffingas            627          627 East Anglia
## 6 Ricberht  Wuffingas            627          630 East Anglia
##    length.of.reign.years reign.was.more.than.30.years
## 1                     NA                           NA
## 2                      7                        FALSE
## 3                     38                         TRUE
## 4                     11                        FALSE
## 5                      0                        FALSE
## 6                      3                        FALSE
```

One of the problems with history is that there is an awful lot of it. Fortunately, odd or messy data can be a really good indicator of the interesting bits of history, so we can narrow it down to the good stuff. For example, although there were seven territories that came together to form England, their boundaries were far from fixed, and sometimes one kingdom would conquer another. We can find these convergences by searching for commas in the `domain` column. To detect a pattern, we use the `str_detect` function. The `fixed` function tells `str_detect` that we are looking for a fixed string (a comma) rather than a regular expression. `str_detect` returns a logical vector that we can use for an index:

```
library(stringr)
multiple_kingdoms <- str_detect(english_monarchs$domain, fixed(","))
english_monarchs[multiple_kingdoms, c("name", "domain")]
```

```
##                             name                    domain
## 17                          Offa       East Anglia, Mercia
## 18                          Offa East Anglia, Kent, Mercia
## 19              Offa and Ecgfrith East Anglia, Kent, Mercia
## 20                       Ecgfrith East Anglia, Kent, Mercia
## 22                       Cœnwulf East Anglia, Kent, Mercia
## 23          Cœnwulf and Cynehelm East Anglia, Kent, Mercia
## 24                       Cœnwulf East Anglia, Kent, Mercia
## 25                       Ceolwulf East Anglia, Kent, Mercia
## 26                      Beornwulf       East Anglia, Mercia
## 82         Ecgbehrt and Æthelwulf             Kent, Wessex
## 83         Ecgbehrt and Æthelwulf     Kent, Mercia, Wessex
## 84         Ecgbehrt and Æthelwulf             Kent, Wessex
## 85        Æthelwulf and Æðelstan I             Kent, Wessex
## 86                      Æthelwulf             Kent, Wessex
## 87 Æthelwulf and Æðelberht III             Kent, Wessex
## 88                Æðelberht III             Kent, Wessex
## 89                    Æthelred I             Kent, Wessex
## 95                        Oswiu     Mercia, Northumbria
```

Similarly, it was quite common for power over a kingdom to be shared between several people, rather than having a single ruler. (This was especially common when a powerful king had several sons.) We can find these instances by looking for either a comma or the word "and" in the name column. This time, since we are looking for two things, it is easier to specify a regular expression rather than a fixed string. The pipe character, |, has the same meaning in regular expressions as it does in R: it means "or."

In this next example, to prevent excessive output we just return the name column and ignore missing values (with is.na):

```
multiple_rulers <- str_detect(english_monarchs$name, ",|and")
english_monarchs$name[multiple_rulers & !is.na(multiple_rulers)]
```

```
##  [1] Sigeberht and Ecgric
##  [2] Hun, Beonna and Alberht
##  [3] Offa and Ecgfrith
##  [4] Cœnwulf and Cynehelm
##  [5] Sighere and Sebbi
##  [6] Sigeheard and Swaefred
##  [7] Eorcenberht and Eormenred
##  [8] Oswine, Swæfbehrt, Swæfheard
##  [9] Swæfbehrt, Swæfheard, Wihtred
## [10] Æðelberht II, Ælfric and Eadberht I
## [11] Æðelberht II and Eardwulf
## [12] Eadberht II, Eanmund and Sigered
## [13] Heaberht and Ecgbehrt II
## [14] Ecgbehrt and Æthelwulf
## [15] Ecgbehrt and Æthelwulf
## [16] Ecgbehrt and Æthelwulf
## [17] Æthelwulf and Æðelstan I
## [18] Æthelwulf and Æðelberht III
## [19] Penda and Eowa
## [20] Penda and Peada
## [21] Æthelred, Lord of the Mercians
## [22] Æthelflæd, Lady of the Mercians
## [23] Ælfwynn, Second Lady of the Mercians
## [24] Hálfdan and Eowils
## [25] Noðhelm and Watt
## [26] Noðhelm and Bryni
## [27] Noðhelm and Osric
## [28] Noðhelm and Æðelstan
## [29] Ælfwald, Oslac and Osmund
## [30] Ælfwald, Ealdwulf, Oslac and Osmund
## [31] Ælfwald, Ealdwulf, Oslac, Osmund and Oswald
## [32] Cenwalh and Seaxburh
## 211 Levels: Adda Æðelbehrt Æðelberht I ... Wulfhere
```

If we wanted to split the name column into individual rulers, then we could use str_split (or strsplit from base R, which does the same thing) in much the same way. str_split accepts a vector and returns a list, since each input string can be split into a vector of possibly differing lengths. If each input must return the same number of splits, we could use `str_split_fixed` instead, which returns a matrix. The output shows the first few examples of multiple rulers:

```
individual_rulers <- str_split(english_monarchs$name, ", | and ")
head(individual_rulers[sapply(individual_rulers, length) > 1])

## [[1]]
## [1] "Sigeberht" "Ecgric"
##
## [[2]]
## [1] "Hun"     "Beonna"  "Alberht"
##
## [[3]]
## [1] "Offa"     "Ecgfrith"
##
## [[4]]
## [1] "Cœnwulf"  "Cynehelm"
##
## [[5]]
## [1] "Sighere" "Sebbi"
##
## [[6]]
## [1] "Sigeheard" "Swaefred"
```

Many of the Anglo-Saxon rulers during this period had Old English characters in their names, like "æ" ("ash"), which represents "ae," or "ð" and "þ" ("eth" and "thorn," respectively), which both represent "th." The exact spelling of each ruler's name isn't standardized in many cases, but to identify a particular ruler it is necessary to be consistent.

Let's take a look at how many times th, ð, and þ are used to form the letters "th." We can count the number of times each one occurs in each name using `str_count`, then `sum` over all rulers to calculate the total number of occurrences:

```
th <- c("th", "ð", "þ")
sapply(          #can also use laply from plyr
  th,
  function(th)
  {
    sum(str_count(english_monarchs$name, th))
  }
)
## th  ð  þ
## 74 26  7
```

It looks like the standard modern Latin spelling is most common in this dataset. If we want to replace the eths and thorns, we can use `str_replace_all`. (A variant function, `str_replace`, replaces only the first match.) Placing eth and thorn in square brackets means "match either of these characters":

```
english_monarchs$new_name <- str_replace_all(english_monarchs$name, "[ðþ]", "th")
```

This sort of trick can be very useful for cleaning up levels of a categorical variable. For example, genders can be specified in several ways in English, but we usually only want two of them. In the next example, we match on a string that starts with (^) "m" and is followed by an optional (?) "ale", which ends the string ($):

```
gender <- c(
  "MALE", "Male", "male", "M", "FEMALE",
  "Female", "female", "f", NA
)
clean_gender <- str_replace(
  gender,
  ignore.case("^m(ale)?$"),
  "Male"
)
(clean_gender <- str_replace(
  clean_gender,
  ignore.case("^f(emale)?$"),
  "Female"
))
## [1] "Male"   "Male"   "Male"   "Male"   "Female" "Female" "Female" "Female"
## [9] NA
```

Manipulating Data Frames

Much of the task of cleaning data involves manipulating data frames to get them into the desired form. We've already seen indexing and the subset function for selecting a subset of a data frame. Other common tasks include augmenting a data frame with additional columns (or replacing existing columns), dealing with missing values, and converting between the wide and long forms of a data frame. There are several functions available for adding or replacing columns in a data frame.

Adding and Replacing Columns

Suppose we want to add a column to the english_monarchs data frame denoting the number of years the rulers were in power. We can use standard assignment to achieve this:

```
english_monarchs$length.of.reign.years <-
  english_monarchs$end.of.reign - english_monarchs$start.of.reign
```

This works, but the repetition of the data frame variable names makes this a lot of effort to type and to read. The with function makes things easier by letting you call variables directly. It takes a data frame[2] and an expression to evaluate:

---

2. Or an environment.

```
english_monarchs$length.of.reign.years <- with(
  english_monarchs,
  end.of.reign - start.of.reign
)
```

The `within` function works in a similar way, but returns the whole data frame:

```
english_monarchs <- within(
  english_monarchs,
  {
    length.of.reign.years <- end.of.reign - start.of.reign
  }
)
```

Although `within` requires more effort in this example, it becomes more useful if you want to change multiple columns:

```
english_monarchs <- within(
  english_monarchs,
  {
    length.of.reign.years <- end.of.reign - start.of.reign
    reign.was.more.than.30.years <- length.of.reign.years > 30
  }
)
```

A good heuristic is that if you are creating or changing one column, then use `with`; if you want to manipulate several columns at once, then use `within`.

An alternative approach is taken by the `mutate` function in the `plyr` package, which accepts new and revised columns as name-value pairs:[3]

```
english_monarchs <- mutate(
  english_monarchs,
  length.of.reign.years       = end.of.reign - start.of.reign,
  reign.was.more.than.30.years = length.of.reign.years > 30
)
```

Dealing with Missing Values

The red deer dataset that we saw in the previous chapter contains measurements of the endocranial volume for each deer using four different techniques. For some but not all of the deer, a second measurement was taken to test the repeatability of the technique. This means that some of the rows have missing values. The `complete.cases` function tells us which rows are free of missing values:

```
data("deer_endocranial_volume", package = "learningr")
has_all_measurements <- complete.cases(deer_endocranial_volume)
deer_endocranial_volume[has_all_measurements, ]
```

---

3. The `transform` function from base R is a precursor to `mutate`, and is now obsolete.

```
##    SkullID VolCT VolBead VolLWH VolFinarelli VolCT2 VolBead2 VolLWH2
## 7     C120   346     335   1250          289    346      330    1264
## 8      C25   302     295   1011          250    303      295    1009
## 9       F7   379     360   1621          347    375      365    1647
## 10     B12   410     400   1740          387    413      395    1728
## 11     B17   405     395   1652          356    408      395    1639
## 12     B18   391     370   1835          419    394      375    1825
## 13      J7   416     405   1834          408    417      405    1876
## 15      A4   336     330   1224          283    345      330    1192
## 20      K2   349     355   1239          286    354      365    1243
```

The `na.omit` function provides a shortcut to this, removing any rows of a data frame where there are missing values:[4]

```
na.omit(deer_endocranial_volume)
```

```
##    SkullID VolCT VolBead VolLWH VolFinarelli VolCT2 VolBead2 VolLWH2
## 7     C120   346     335   1250          289    346      330    1264
## 8      C25   302     295   1011          250    303      295    1009
## 9       F7   379     360   1621          347    375      365    1647
## 10     B12   410     400   1740          387    413      395    1728
## 11     B17   405     395   1652          356    408      395    1639
## 12     B18   391     370   1835          419    394      375    1825
## 13      J7   416     405   1834          408    417      405    1876
## 15      A4   336     330   1224          283    345      330    1192
## 20      K2   349     355   1239          286    354      365    1243
```

By contrast, `na.fail` will throw an error if your data frame contains any missing values:

```
na.fail(deer_endocranial_volume)
```

Both these functions can accept vectors as well, removing missing values or failing, as in the data frame case.

> You can use multiple imputation to fill in missing values in a statistically sound way. This is beyond the scope of the book, but the `mice` and `mix` packages are good places to start.

Converting Between Wide and Long Form

The red deer dataset contains measurements of the volume of deer skulls obtained in four different ways. Each measurement for a particular deer is given in its own column. (For simplicity, let's ignore the columns for repeat measurements.) This is known as the *wide* form of a data frame:

```
deer_wide <- deer_endocranial_volume[, 1:5]
```

---

4. `na.exclude` does the same thing as `na.omit`; their dual existence is mostly for legacy purposes.

An alternative point of view is that each skull measurement is the same type of thing (a measurement), just a different measurement. So, a different way of representing the data would be to have four rows for each deer, with a column for the skull ID, as before (so each value would be repeated four times), a column containing all the measurements, and a factor column explaining what type of measurement is contained in that particular row. This is called the *long* form of a data frame.

There is a function in base R for converting between wide and long form, called reshape. It's very powerful, but not entirely intuitive; a better alternative is to use the functionality of the reshape2 package.

The melt function available in this package converts from wide form to long.[5] We choose SkullID as the ID column (with everything else being classed as a measurement):

```
library(reshape2)
deer_long <- melt(deer_wide, id.vars = "SkullID")
head(deer_long)

##     SkullID variable value
## 1    DIC44    VolCT   389
## 2      B11    VolCT   389
## 3    DIC90    VolCT   352
## 4    DIC83    VolCT   388
## 5   DIC787    VolCT   375
## 6  DIC1573    VolCT   325
```

You can, alternatively, supply the measure.vars argument, which is all the columns that aren't included in id.vars. In this case it is more work, but it can be useful if you have many ID variables and few measurement variables:

```
melt(deer_wide, measure.vars = c("VolCT", "VolBead", "VolLWH", "VolFinarelli"))
```

The dcast function converts back from long to wide and returns the result as a data frame (the related function acast returns a vector, matrix, or array):

```
deer_wide_again <- dcast(deer_long, SkullID ~ variable)
```

Our reconstituted dataset, deer_wide_again, is identical to the original, deer_wide, except that it is now ordered alphabetically by SkullID.

 Spreadsheet aficionados might note that acast and dcast are effectively creating pivot tables.

---

5. The terminology of "melting" and its opposite, "casting," is borrowed from the steel industry.

Using SQL

The `sqldf` package provides a way of manipulating data frames using SQL. In general, native R functions are more concise and readable than SQL code, but if you come from a database background this package can ease your transition to R:

```
install.packages("sqldf")
```

The next example compares the native R and `sqldf` versions of a subsetting query:

```
library(sqldf)
## Loading required package: DBI
## Loading required package: gsubfn
## Loading required package: proto
## Loading required namespace: tcltk
## Loading required package: chron
## Loading required package: RSQLite
## Loading required package: RSQLite.extfuns
subset(
  deer_endocranial_volume,
  VolCT > 400 | VolCT2 > 400,
  c(VolCT, VolCT2)
)
##     VolCT VolCT2
## 10    410    413
## 11    405    408
## 13    416    417
## 16    418     NA
query <-
  "SELECT
      VolCT,
      VolCT2
    FROM
      deer_endocranial_volume
    WHERE
      VolCT > 400 OR
      VolCT2 > 400"
sqldf(query)
## Loading required package: tcltk
##    VolCT VolCT2
## 1    410    413
## 2    405    408
## 3    416    417
## 4    418     NA
```

Sorting

It's often useful to have numeric data in size order, since the interesting values are often the extremes. The sort function sorts vectors from smallest to largest (or largest to smallest):[6]

```
x <- c(2, 32, 4, 16, 8)
sort(x)
```

```
## [1]  2  4  8 16 32
```

```
sort(x, decreasing = TRUE)
```

```
## [1] 32 16  8  4  2
```

Strings can also be sorted, but the sort order depends upon locale. Usually letters are ordered from "a" through to "z," but there are oddities: in Estonian, "z" comes after "s" and before "t," for example. More of these quirks are listed in the ?Comparison help page. In an English or North American locale, you'll see results like this:

```
sort(c("I", "shot", "the", "city", "sheriff"))
```

```
## [1] "city"    "I"       "sheriff" "shot"    "the"
```

The order function is a kind of inverse to sort. The *i*th element of the order contains the index of the element of x that will end up in the *i*th position after sorting. That takes a bit of getting your head around, but mostly what you need to know is that x[order(x)] returns the same result as sort(x):

```
order(x)
```

```
## [1] 1 3 5 4 2
```

```
x[order(x)]
```

```
## [1]  2  4  8 16 32
```

```
identical(sort(x), x[order(x)])
```

```
## [1] TRUE
```

order is most useful for sorting data frames, where sort cannot be used directly. For example, to sort the english_monarchs data frame by the year of the start of the reign, we can use:

```
year_order <- order(english_monarchs$start.of.reign)
english_monarchs[year_order, ]
```

The arrange function from the plyr package provides a one-line alternative for ordering data frames:

---

6. Sorting algorithm geeks may like to note that the sort function uses a shellsort by default, but you can use method = "quick" to use a quicksort instead. Radix sorting is also available for factors.

```
arrange(english_monarchs, start.of.reign)
```

The rank function gives the rank of each element in a dataset, providing a few ways of dealing with ties:

```
(x <- sample(3, 7, replace = TRUE))
## [1] 1 2 1 3 3 3 2
rank(x)
## [1] 1.5 3.5 1.5 6.0 6.0 6.0 3.5
rank(x, ties.method = "first")
## [1] 1 3 2 5 6 7 4
```

Functional Programming

Several concepts from functional programming languages like LISP and Haskell have been introduced into R. You don't need to know anything at all about functional programming to use them;[7] you just need to know that these functions can be useful for manipulating data.

The Negate function accepts a predicate (that is, a function that returns a logical vector), and returns another predicate that does the opposite.[8] It returns TRUE when the input returns FALSE and FALSE when the input returns TRUE:

```
ct2 <- deer_endocranial_volume$VolCT2  #for convenience of typing
isnt.na <- Negate(is.na)
identical(isnt.na(ct2), !is.na(ct2))
## [1] TRUE
```

Filter takes a function that returns a logical vector and an input vector, and returns only those values where the function returns TRUE:

```
Filter(isnt.na, ct2)
## [1] 346 303 375 413 408 394 417 345 354
```

The Position function behaves a little bit like which, which we saw in "Vectors" on page 39 in Chapter 4. It returns the first index where applying a predicate to a vector returns TRUE:

```
Position(isnt.na, ct2)
## [1] 7
```

---

7. For the curious, there's a good introduction at wordIQ.com (*http://bit.ly/17mnkKs*).

8. Technically, a predicate is a function that returns a single logical value, so we're abusing the term, but the word for the vector equivalent hasn't been coined yet. I rather like the Schwarzenegger-esque "predicator," but until that catches on, we'll use plain old predicate.

Find is similar to `Position`, but it returns the first value rather than the first index:

```
Find(isnt.na, ct2)
## [1] 346
```

Map applies a function element-wise to its inputs. It's just a wrapper to `mapply`, with `SIMPLIFY = FALSE`. In this next example, we retrieve the average measurement using each method for each deer in the red deer dataset. First, we need a function to pass to Map to find the volume of each deer skull:

```
get_volume <- function(ct, bead, lwh, finarelli, ct2, bead2, lwh2)
{
  #If there is a second measurement, take the average
  if(!is.na(ct2))
  {
    ct <- (ct + ct2) / 2
    bead <- (bead + bead2) / 2
    lwh <- (lwh + lwh2) / 2
  }
  #Divide lwh by 4 to bring it in line with other measurements
  c(ct = ct, bead = bead, lwh.4 = lwh / 4, finarelli = finarelli)
}
```

Then Map behaves like `mapply`—it takes a function and then each argument to pass to that function:

```
measurements_by_deer <- with(
  deer_endocranial_volume,
  Map(
    get_volume,
    VolCT,
    VolBead,
    VolLWH,
    VolFinarelli,
    VolCT2,
    VolBead2,
    VolLWH2
  )
)
head(measurements_by_deer)

## [[1]]
##        ct      bead     lwh.4 finarelli
##       389       375       371       337
##
## [[2]]
##        ct      bead     lwh.4 finarelli
##     389.0     370.0     430.5     377.0
##
## [[3]]
##        ct      bead     lwh.4 finarelli
##     352.0     345.0     373.8     328.0
```

```
##
## [[4]]
##          ct       bead      lwh.4 finarelli
##        388.0     370.0      420.8    377.0
##
## [[5]]
##          ct       bead      lwh.4 finarelli
##        375.0     355.0      364.5    328.0
##
## [[6]]
##          ct       bead      lwh.4 finarelli
##        325.0     320.0      340.8    291.0
```

The Reduce function turns a binary function into one that accepts multiple inputs. For example, the + operator calculates the sum of two numbers, but the sum function calculates the sum of multiple inputs. sum(a, b, c, d, e) is (roughly) equivalent to Reduce("+", list(a, b, c, d, e)).

We can define a simple binary function that calculates the (parallel) maximum of two inputs:

```
pmax2 <- function(x, y) ifelse(x >= y, x, y)
```

If we reduce this function, then it will accept a list of many inputs (like the pmax function in base R does):

```
Reduce(pmax2, measurements_by_deer)

##          ct       bead      lwh.4 finarelli
##        418.0     405.0      463.8    419.0
```

One proviso is that Reduce repeatedly calls the binary function on pairs of inputs, so:

```
Reduce("+", list(a, b, c, d, e))
```

is the same as:

```
((((a + b) + c) + d) + e)
```

This means that you can't use it for something like calculating the mean, since:

```
mean(mean(mean(mean(a, b), c), d), e) != mean(a, b, c, d, e)
```

Summary

- The stringr package is useful for manipulating strings.
- Columns of a data frame can be added, subtracted, or manipulated.
- Data frames can exist in *wide* or *long* form.
- Vectors can be sorted, ranked, and ordered.

- R has some functional programming capabilities, including `Map` and `Reduce`.

Test Your Knowledge: Quiz

Question 13-1
> How would you count the number of times the word "thou" appears in Shakespeare's *The Tempest*?

Question 13-2
> Name as many functions as you can think of for adding columns to data frames.

Question 13-3
> What is the opposite of melting?

Question 13-4
> How would you reorder a data frame by one if its columns?

Question 13-5
> How would you find the first positive number in a vector?

Test Your Knowledge: Exercises

Exercise 13-1
> 1. Load the `hafu` dataset from the `learningr` package. In the `Father` and `Mother` columns, some values have question marks after the country name, indicating that the author was uncertain about the nationality of the parent. Create two new columns in the `hafu` data frame, denoting whether or not there was a question mark in the `Father` or `Mother` column, respectively.
>
> 2. Remove those question marks from the `Father` and `Mother` columns. [10]

Exercise 13-2
> The `hafu` dataset has separate columns for the nationality of each parent. Convert the data frame from wide form to long form, with a single column for the parents' nationality and a column indicating which parent the nationality refers to. [5]

Exercise 13-3
> Write a function that returns the 10 most common values in a vector, along with their counts. Try the function on some columns from the `hafu` dataset. [10]

## CHAPTER 14

Exploring and Visualizing

Once you've imported your data and cleaned and transformed it into a suitable state, you get to start asking questions like "what does it all mean?" The two main tools at your disposal are summary statistics and plots. (Modeling comes later, because you need to understand your data before you can model it properly.) R is well served by a comprehensive set of functions for calculating statistics, and a choice of three different graphics systems.

Chapter Goals

After reading this chapter, you should:

- Be able to calculate a range of summary statistics on numeric data
- Be able to draw standard plots in R's three plotting systems
- Be able to manipulate those plots in simple ways

Summary Statistics

We've already come across many of the functions for calculating summary statistics, so this section is partly a recap. Most are fairly obvious in their naming and their usage; for example, `mean` and `median` calculate their respective measures of location. There isn't a function for the mode, but it can be calculated from the results of the `table` function, which gives counts of each element. (If you haven't already, have a go at Exercise 13-3 now.)

In the following examples, the `obama_vs_mccain` dataset contains the fractions of people voting for Obama and McCain in the 2008 US presidential elections, along with some contextual background information on demographics:

```
data(obama_vs_mccain, package = "learningr")
obama <- obama_vs_mccain$Obama
mean(obama)
```

```
## [1] 51.29
```

```
median(obama)
```

```
## [1] 51.38
```

The table function doesn't make a great deal of sense for the obama variable (or many numeric variables) since each value is unique. By combining it with cut, we can see how many values fall into different bins:

```
table(cut(obama, seq.int(0, 100, 10)))
```

```
##
##    (0,10]   (10,20]   (20,30]   (30,40]   (40,50]   (50,60]   (60,70]   (70,80]
##        0         0         0         8        16        16         9         1
##   (80,90]  (90,100]
##        0         1
```

var and sd calculate the variance and standard deviation, respectively. Slightly less common is the mad function for calculating the mean absolute deviation:

```
var(obama)
```

```
## [1] 123.1
```

```
sd(obama)
```

```
## [1] 11.09
```

```
mad(obama)
```

```
## [1] 11.49
```

There are several functions for getting the extremes of numeric data. min and max are the most obvious, giving the smallest and largest values of all their inputs, respectively. pmin and pmax (the "parallel" equivalents) calculate the smallest and largest values at each point across several vectors of the same length. Meanwhile, the range function gives the minimum and maximum in a single function call:

```
min(obama)
```

```
## [1] 32.54
```

```
with(obama_vs_mccain, pmin(Obama, McCain))
```

```
##  [1] 38.74 37.89 44.91 38.86 36.91 44.71 38.22  6.53 36.93 48.10 46.90
## [12] 26.58 35.91 36.74 48.82 44.39 41.55 41.15 39.93 40.38 36.47 35.99
## [23] 40.89 43.82 43.00 49.23 47.11 41.60 42.65 44.52 41.61 41.78 36.03
## [34] 49.38 44.50 46.80 34.35 40.40 44.15 35.06 44.90 44.75 41.79 43.63
## [45] 34.22 30.45 46.33 40.26 42.51 42.31 32.54
```

```
range(obama)
```

```
## [1] 32.54 92.46
```

cummin and cummax provide the smallest and largest values so far in a vector. Similarly, cumsum and cumprod provide sums and products of the values to date. These functions make most sense when the input has been ordered in a useful way:

```
cummin(obama)
```

```
## [1] 38.74 37.89 37.89 37.89 37.89 37.89 37.89 37.89 37.89 37.89 37.89
## [12] 37.89 35.91 35.91 35.91 35.91 35.91 35.91 35.91 35.91 35.91
## [23] 35.91 35.91 35.91 35.91 35.91 35.91 35.91 35.91 35.91 35.91 35.91
## [34] 35.91 35.91 35.91 34.35 34.35 34.35 34.35 34.35 34.35 34.35 34.35
## [45] 34.22 34.22 34.22 34.22 34.22 34.22 32.54
```

```
cumsum(obama)
```

```
## [1]   38.74   76.63  121.54  160.40  221.34  275.00  335.59  428.05
## [9]  489.96  540.87  587.77  659.62  695.53  757.38  807.23  861.16
## [17]  902.71  943.86  983.79 1041.50 1103.42 1165.22 1222.55 1276.61
## [25] 1319.61 1368.84 1415.95 1457.55 1512.70 1566.83 1623.97 1680.88
## [33] 1743.76 1793.46 1837.96 1889.34 1923.69 1980.44 2034.91 2097.77
## [41] 2142.67 2187.42 2229.21 2272.84 2307.06 2374.52 2427.15 2484.49
## [49] 2527.00 2583.22 2615.76
```

```
cumprod(obama)
```

```
## [1] 3.874e+01 1.468e+03 6.592e+04 2.562e+06 1.561e+08 8.377e+09 5.076e+11
## [8] 4.693e+13 2.905e+15 1.479e+17 6.937e+18 4.984e+20 1.790e+22 1.107e+24
## [15] 5.519e+25 2.976e+27 1.237e+29 5.089e+30 2.032e+32 1.173e+34 7.261e+35
## [22] 4.487e+37 2.572e+39 1.391e+41 5.980e+42 2.944e+44 1.387e+46 5.769e+47
## [29] 3.182e+49 1.722e+51 9.841e+52 5.601e+54 3.522e+56 1.750e+58 7.789e+59
## [36] 4.002e+61 1.375e+63 7.801e+64 4.249e+66 2.671e+68 1.199e+70 5.367e+71
## [43] 2.243e+73 9.785e+74 3.349e+76 2.259e+78 1.189e+80 6.817e+81 2.898e+83
## [50] 1.629e+85 5.302e+86
```

The quantile function provides, as you might expect, quantiles (median, min, and max are special cases). It defaults to the median, minimum, maximum, and lower and upper quartiles, and in an impressive feat of overengineering, it gives a choice of nine different calculation algorithms:

```
quantile(obama)
```

```
##    0%   25%   50%   75%  100%
## 32.54 42.75 51.38 57.34 92.46
```

```
quantile(obama, type = 5)     #to reproduce SAS results
```

```
##    0%   25%   50%   75%  100%
## 32.54 42.63 51.38 57.34 92.46
```

```
quantile(obama, c(0.9, 0.95, 0.99))
```

```
##   90%   95%   99%
## 61.92 65.17 82.16
```

IQR wraps quantile to give the interquartile range (the 75th percentile minus the 25th percentile):

```
IQR(obama)
```

```
## [1] 14.58
```

fivenum provides a faster, greatly simplified alternative to quantile. You only get one algorithm, and only the default quantiles can be calculated. It has a niche use where speed matters:

```
fivenum(obama)
```

```
## [1] 32.54 42.75 51.38 57.34 92.46
```

There are some shortcuts for calculating multiple statistics at once. You've already met the summary function, which accepts vectors or data frames:

```
summary(obama_vs_mccain)
```

```
##       State              Region        Obama          McCain
## Alabama    : 1   IV     : 8   Min.   :32.5   Min.   : 6.53
## Alaska     : 1   I      : 6   1st Qu.:42.8   1st Qu.:40.39
## Arizona    : 1   III    : 6   Median :51.4   Median :46.80
## Arkansas   : 1   V      : 6   Mean   :51.3   Mean   :47.00
## California : 1   VIII   : 6   3rd Qu.:57.3   3rd Qu.:55.88
## Colorado   : 1   VI     : 5   Max.   :92.5   Max.   :65.65
## (Other)    :45   (Other):14
##     Turnout        Unemployment       Income         Population
## Min.   :50.8   Min.   :3.40   Min.   :19534   Min.   :  563626
## 1st Qu.:61.0   1st Qu.:5.05   1st Qu.:23501   1st Qu.: 1702662
## Median :64.9   Median :5.90   Median :25203   Median : 4350606
## Mean   :64.1   Mean   :6.01   Mean   :26580   Mean   : 6074128
## 3rd Qu.:68.0   3rd Qu.:7.25   3rd Qu.:28978   3rd Qu.: 6656506
## Max.   :78.0   Max.   :9.40   Max.   :40846   Max.   :37341989
## NA's   :4
##     Catholic        Protestant        Other       Non.religious     Black
## Min.   : 6.0   Min.   :26.0   Min.   :0.00   Min.   : 5   Min.   : 0.4
## 1st Qu.:12.0   1st Qu.:46.0   1st Qu.:2.00   1st Qu.:12   1st Qu.: 3.1
## Median :21.0   Median :54.0   Median :3.00   Median :15   Median : 7.4
## Mean   :21.7   Mean   :53.8   Mean   :3.29   Mean   :16   Mean   :11.1
## 3rd Qu.:29.0   3rd Qu.:62.0   3rd Qu.:4.00   3rd Qu.:19   3rd Qu.:15.2
## Max.   :46.0   Max.   :80.0   Max.   :8.00   Max.   :34   Max.   :50.7
## NA's   :2      NA's   :2      NA's   :2      NA's   :2
##     Latino        Urbanization
## Min.   : 1.2   Min.   :   1
## 1st Qu.: 4.3   1st Qu.:  46
## Median : 8.2   Median : 101
## Mean   :10.3   Mean   : 386
## 3rd Qu.:12.1   3rd Qu.: 221
## Max.   :46.3   Max.   :9856
##
```

The cor function calculates correlations between numeric vectors. As you would expect, there was an almost perfect negative correlation between the fraction of people voting for Obama and the fraction of people voting for McCain. (The slight imperfection is caused by voters for independent candidates.) The cancor function (short for "canonical correlation") provides extra details, and the cov function calculates covariances:

```
with(obama_vs_mccain, cor(Obama, McCain))
```

```
## [1] -0.9981
```

```
with(obama_vs_mccain, cancor(Obama, McCain))
```

```
## $cor
## [1] 0.9981
##
## $xcoef
##          [,1]
## [1,] 0.01275
##
## $ycoef
##          [,1]
## [1,] -0.01287
##
## $xcenter
## [1] 51.29
##
## $ycenter
## [1] 47
```

```
with(obama_vs_mccain, cov(Obama, McCain))
```

```
## [1] -121.7
```

The Three Plotting Systems

Over its lifetime, R has accumulated three different plotting systems. base graphics are the oldest system, having been around as long as R itself. base graphs are easy to get started with, but they require a lot of fiddling and magic incantations to polish, and are very hard to extend to new graph types.

To remedy some of the limitations of base, the grid graphics system was developed to allow more flexible plotting. grid lets you draw things at a very low level, specifying where to draw each point, line, or rectangle. While this is wonderful, none of us have time to write a couple of hundred lines of code each time we want to draw a scatterplot.

The second plotting system, lattice, is built on top of the grid system, providing high-level functions for all the common plot types. It has two standout features that aren't available in base graphics. First, the results of each plot are saved into a variable, rather than just being drawn on the screen. This means that you can draw something, edit it, and draw it again; groups of related plots are easier to draw, and plots can be saved between sessions. The second great feature is that plots can contain multiple panels in a lattice,[1] so you can split up your data into categories and compare the differences between groups. This solves the plotting equivalent of the split-apply-combine problem that we discussed in Chapter 9.

The ggplot2 system, also built on top of grid, is the most modern of the three plotting systems. The "gg" stands for "grammar of graphics,"[2] which aims to break down graphs into component chunks. The result is that code for a ggplot looks a bit like the English way of articulating what you want in the graph.

The three systems are, sadly, mostly incompatible (there are ways to combine base and grid graphics, but they should be considered a last resort). The good news is that you can do almost everything you want in ggplot2, so learning all three systems is mostly overkill. There are a couple of rare use cases where ggplot2 isn't appropriate: it does more calculation than other graphics systems, so for quick and dirty plots of very large datasets it can be more convenient to use another system. Also, many plotting packages are based on one of the other two systems, so using those packages requires a little knowledge of base or lattice.

The following examples demonstrate all three systems; if you are pushed for time, then just take note of the ggplot2 parts. Due to space constraints, this chapter can only give a taste of some of the possibilities on offer. Fortunately, there are three excellent and easy to read books on graph drawing in R, namely R Graphics, ggplot2, and Lattice, by the authors of the grid, ggplot2, and lattice systems, respectively.[3]

Scatterplots

Perhaps the most common of all plots is the scatterplot, used for exploring the relationships between two continuous variables. The obama_vs_mccain dataset has lots of numeric variables that we can compare, but we'll start by asking, "Does voter income affect turnout at the polls?"

---

1. There are several terms for this. Edward Tufte called the idea "small multiples" in *Envisioning Information*; Bill Cleveland and Rick Becker of Bell Labs coined the term "trellising"; Deepayan Sarkar renamed it "latticing" in the lattice package to avoid a Bell Labs trademark; and Leland Wilkinson named it "faceting," a term that is used in ggplot2.

2. The concept was devised by Leland Wilkinson in the book of the same name. The book is brilliant, but not suitable for bedtime reading, being densely packed with equations.

3. Books on plotting have the advantage that even if you can't be bothered to read them, they are pretty to flick through.

Take 1: base Graphics

The base graphic function to draw a scatterplot is simply plot. The best-practice code style these days is to keep all the variables you want for a plot together inside a data frame (or possibly several), rather than having them scattered in individual vectors. Unfortunately, plot predates[4] this idea, so we have to wrap it in a call to with to access the columns.

Although plot will simply ignore missing values, for tidiness let's remove the rows with missing Turnout values:

```
obama_vs_mccain <- obama_vs_mccain[!is.na(obama_vs_mccain$Turnout), ]
```

We can then create a simple scatterplot, shown in Figure 14-1:

```
with(obama_vs_mccain, plot(Income, Turnout))
```

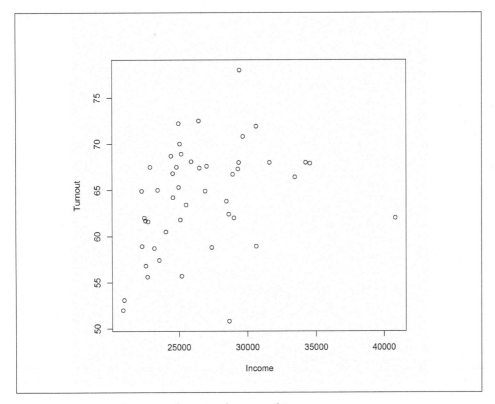

Figure 14-1. A simple scatterplot using base graphics

4. Predates as in "comes before," not "hunts and eats."

plot has many arguments for customizing the output, some of which are more intuitive than others. col changes the color of the points. It accepts any of the named colors returned by colors, or an HTML-style hex value like "#123456". You can change the shape of the points with the pch argument (short for "plot character").[5] Figure 14-2 shows an updated scatterplot, changing the point color to violet and the point shape to filled-in circles:

```
with(obama_vs_mccain, plot(Income, Turnout, col = "violet", pch = 20))
```

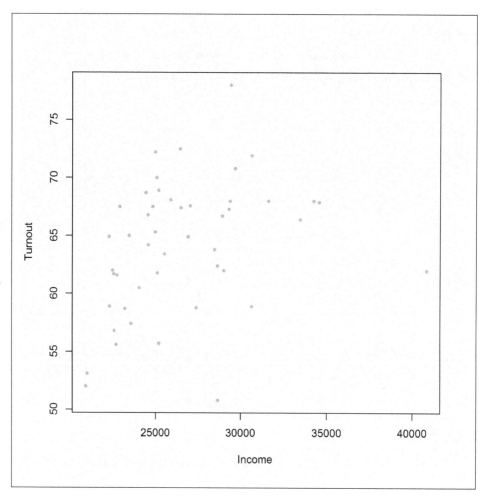

Figure 14-2. Setting color and point shape using base graphics

5. Read the ?points help page and try plot(1:25, pch = 1:25, bg = "blue") to see the different shapes.

Log scales are possible by setting the log argument. log = "x" means use a logarithmic x-scale, log = "y" means use a logarithmic y-scale, and log = "xy" makes both scales logarithmic. Figures 14-3 and 14-4 display some options for log-scaled axes:

```
with(obama_vs_mccain, plot(Income, Turnout, log = "y"))
#Fig. 14-3

with(obama_vs_mccain, plot(Income, Turnout, log = "xy"))
#Fig. 14-4
```

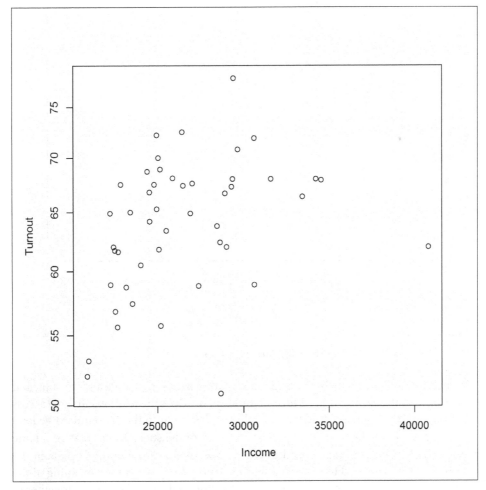

Figure 14-3. Log y-scale using base graphics

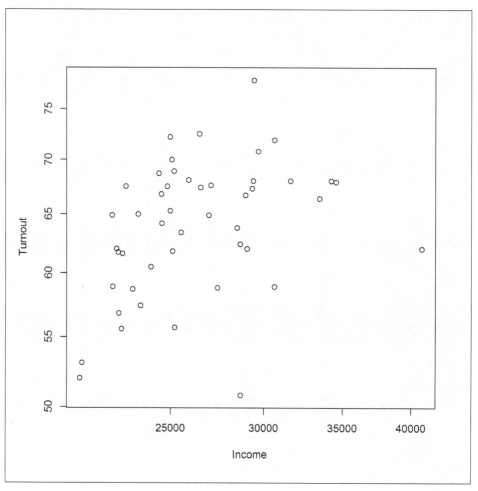

Figure 14-4. Log x- and y-scales using base graphics

We can see that there is a definite positive correlation between income and turnout, and it's stronger on the log-log scale. A further question is, "Does the relationship hold across all of the USA?" To answer this, we can split the data up into the 10 Standard Federal Regions given in the `Region` column, and plot each of the subsets in a "matrix" in one figure. The `layout` function is used to control the layout of the multiple plots in the matrix. Don't feel obliged to spend a long time trying to figure out the meaning of the next code chunk; it only serves to show that drawing multiple related plots together in `base` graphics is possible. Sadly, the code invariably looks like it fell out of the proverbial ugly tree, so this technique should only be used as a last resort. Figure 14-5 shows the result:

```
par(mar = c(3, 3, 0.5, 0.5), oma = rep.int(0, 4), mgp = c(2, 1, 0))
regions <- levels(obama_vs_mccain$Region)
plot_numbers <- seq_along(regions)
layout(matrix(plot_numbers, ncol = 5, byrow = TRUE))
for(region in regions)
{
  regional_data <- subset(obama_vs_mccain, Region == region)
  with(regional_data,  plot(Income, Turnout))
}
```

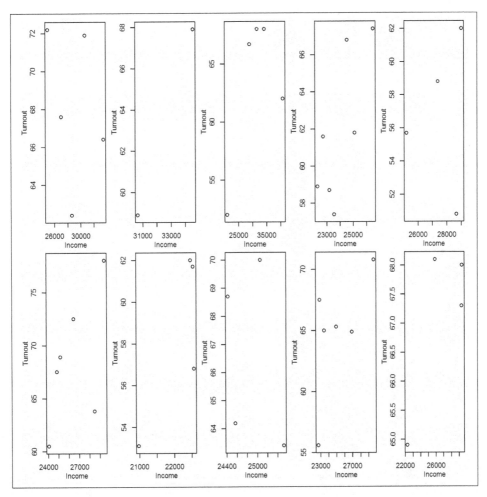

Figure 14-5. Multiple plots in the same figure using base graphics

Take 2: lattice Graphics

The lattice equivalent of plot is xyplot. It uses a formula interface to specify the variables for the *x* and *y* coordinates. Formulae will be discussed in more depth in "Formulae" on page 267, but for now just note that you need to type yvar ~ xvar. Conveniently, xyplot (and other lattice functions) takes a data argument that tells it which data frame to look for variables in. Figure 14-6 shows the lattice equivalent of Figure 14-1:

```
library(lattice)
xyplot(Turnout ~ Income, obama_vs_mccain)
```

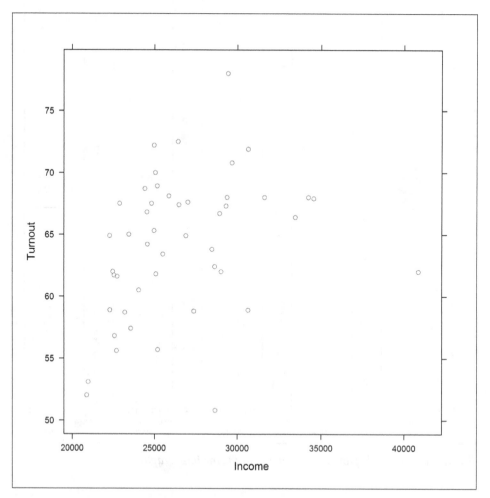

Figure 14-6. A simple scatterplot using lattice graphics

Many of the options for changing plot features are the same as those in base graphics. Figure 14-7 changes the color and point shape, mimicking Figure 14-2:

```
xyplot(Turnout ~ Income, obama_vs_mccain, col = "violet", pch = 20)
```

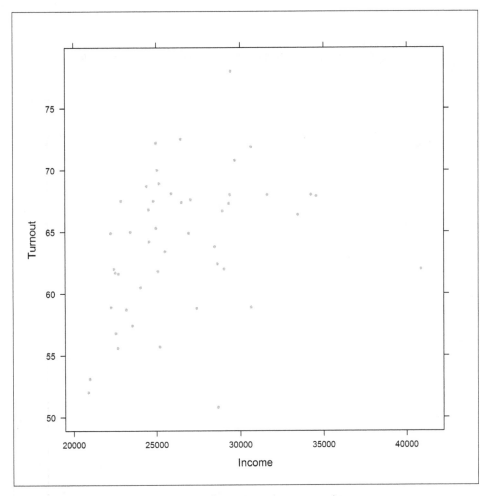

Figure 14-7. Setting color and point shape using lattice graphics

Axis scales, however, are specified in a different way. lattice plots take a scales argument, which must be a list. The contents of this list must be *name* = *value* pairs; for example, log = TRUE sets a log scale for both axes. The scales list can also take further (sub)list arguments named x and y that specify settings for only those axes. Don't panic, it isn't as complicated as it sounds. Figures 14-8 and 14-9 show examples of scaled axes:

```
xyplot(
  Turnout ~ Income,
  obama_vs_mccain,
  scales = list(log = TRUE)              #both axes log scaled (Fig. 14-8)
)

xyplot(
  Turnout ~ Income,
  obama_vs_mccain,
  scales = list(y = list(log = TRUE))  #y-axis log scaled (Fig. 14-9)
)
```

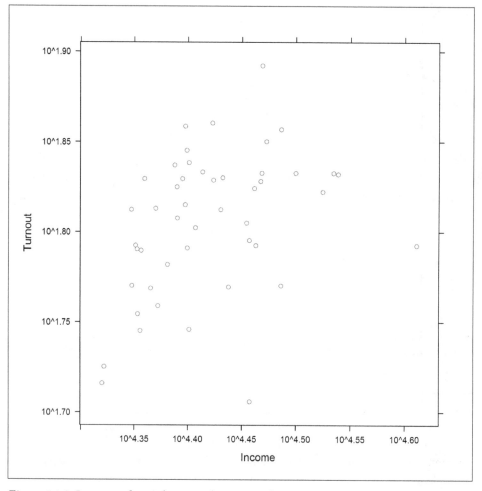

Figure 14-8. Log x- and y-scales using lattice graphics

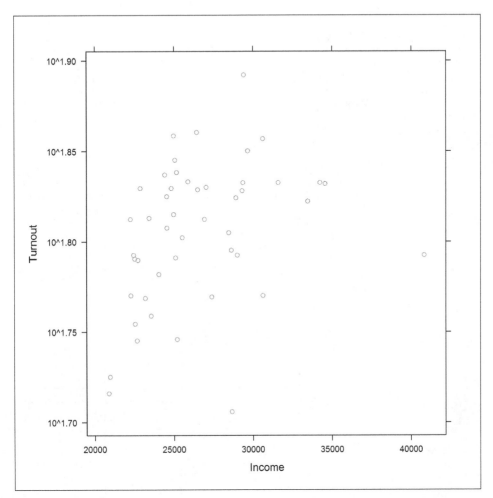

Figure 14-9. Log y-scale using lattice graphics

The formula interface makes splitting the data by region vastly easier. All we have to do is to append a | (that's a "pipe" character; the same one that is used for logical "or") and the variable that we want to split by, in this case `Region`. Using the argument `relation` = `"same"` means that each panel shares the same axes. Axis ticks for each panel are drawn on alternating sides of the plot when the `alternating` argument is `TRUE` (the default), or just the left and bottom otherwise. The output is shown in Figure 14-10; notice the improvement over Figure 14-5:

```
xyplot(
  Turnout ~ Income | Region,
  obama_vs_mccain,
  scales = list(
    log        = TRUE,
    relation   = "same",
    alternating = FALSE
  ),
  layout = c(5, 2)
)
```

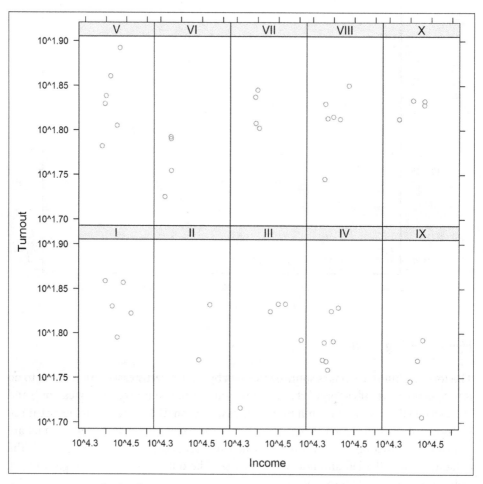

Figure 14-10. Multiple plots in the same figure using lattice graphics

Another benefit is that `lattice` plots are stored in variables, (as opposed to `base` plots, which are just drawn in a window) so we can sequentially update them. Figure 14-11 shows a `lattice` plot that is updated in Figure 14-12:

```
(lat1 <- xyplot(
  Turnout ~ Income | Region,
  obama_vs_mccain
))
#Fig. 14-11

(lat2 <- update(lat1, col = "violet", pch = 20))
#Fig. 14-12
```

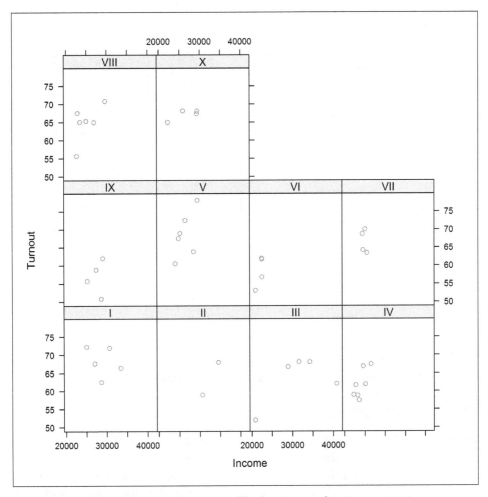

Figure 14-11. This plot is stored as a variable that is reused in Figure 14-12

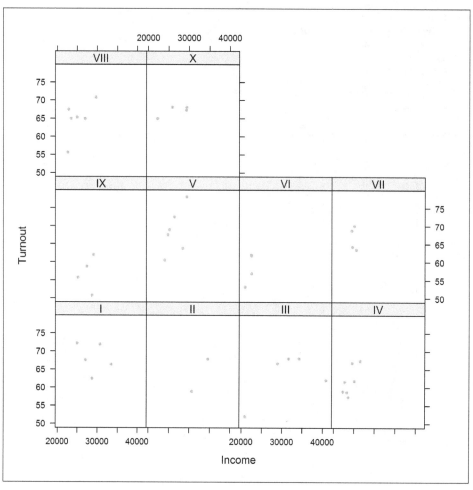

Figure 14-12. This plot reuses a lattice variable from Figure 14-11

Take 3: ggplot2 Graphics

ggplot2 (the "2" is because it took a couple of attempts to get it right) takes many of the good ideas in lattice and builds on them. So, splitting plots up into panels is easy, and sequentially building plots is also possible. Beyond that, ggplot2 has a few special tricks of its own. Most importantly, its "grammatical" nature means that it consists of small building blocks, so it's easier to create brand new plot types, if you feel so inclined.

The syntax is a very different to other plotting code, so mentally prepare yourself to look at something new. Each plot is constructed with a call to the ggplot function, which takes a data frame as its first argument and an *aesthetic* as its second. In practice, that means passing the columns for the x and y variables to the `aes` function. We then add a *geom* to tell the plot to display some points. Figure 14-13 shows the result:

```
library(ggplot2)
ggplot(obama_vs_mccain, aes(Income, Turnout)) +
  geom_point()
```

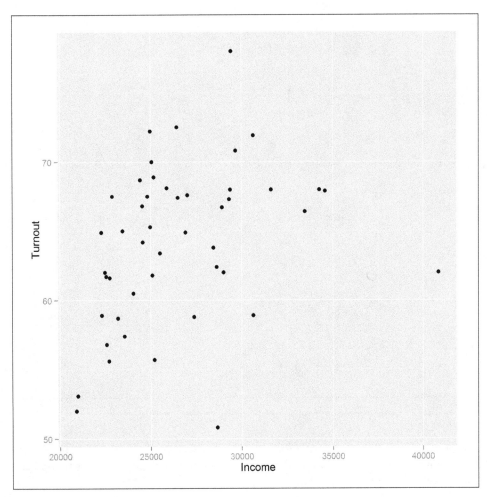

Figure 14-13. A simple scatterplot using ggplot2 graphics

ggplot2 recognizes the commands from `base` for changing the color and shape of the points, but also has its own set of more human-readable names. In Figure 14-14, "shape" replaces "pch," and color can be specified using either "color" or "colour":

```
ggplot(obama_vs_mccain, aes(Income, Turnout)) +
  geom_point(color = "violet", shape = 20)
```

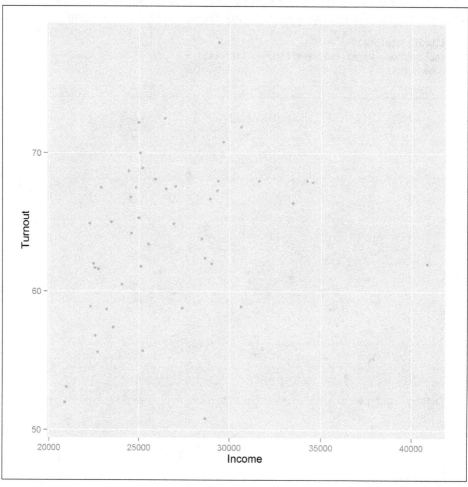

Figure 14-14. Setting color and point shape using ggplot2 graphics

To set a log scale, we add a *scale* for each axis, as seen in Figure 14-15. The breaks argument specifies the locations of the axis ticks. It is optional, but used here to replicate the behavior of the base and +lattice examples:

```
ggplot(obama_vs_mccain, aes(Income, Turnout)) +
  geom_point() +
  scale_x_log10(breaks = seq(2e4, 4e4, 1e4)) +
  scale_y_log10(breaks = seq(50, 75, 5))
```

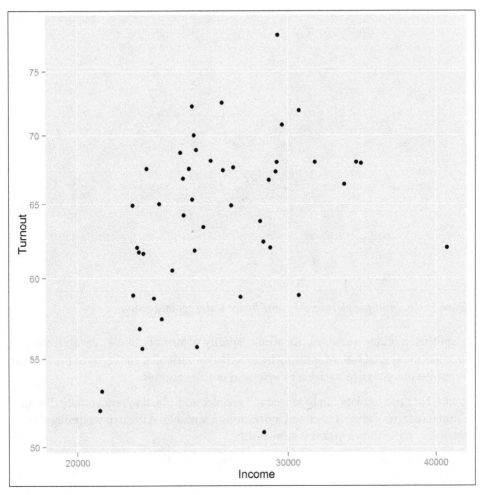

Figure 14-15. Log scales using ggplot2 graphics

To split the plot into individual panels, we add a *facet*. Like the lattice plots, facets take a formula argument. Figure 14-16 demonstrates the facet_wrap function. For easy reading, the x-axis ticks have been rotated by 30 degrees and right-justified using the theme function:

```
ggplot(obama_vs_mccain, aes(Income, Turnout)) +
  geom_point() +
  scale_x_log10(breaks = seq(2e4, 4e4, 1e4)) +
  scale_y_log10(breaks = seq(50, 75, 5)) +
  facet_wrap(~ Region, ncol = 4)
```

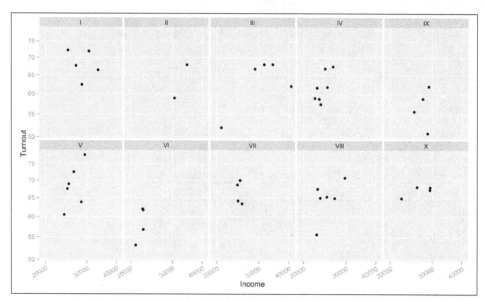

Figure 14-16. Multiple plots in the same figure using ggplot2 graphics

To split by multiple variables, we would specify a formula like ~ var1 + var2 + var3. For the special case of splitting by exactly two variables, facet_grid provides an alternative that puts one variable in rows and one in columns.

As with lattice, ggplots can be stored in variables and added to sequentially. The next example redraws Figure 14-13 and stores it as a variable. As usual, wrapping the expression in parentheses makes it auto-print:

```
(gg1 <- ggplot(obama_vs_mccain, aes(Income, Turnout)) +
  geom_point()
)
```

Figure 14-17 shows the output. We can then update it as follows, with the result shown in Figure 14-18:

```
(gg2 <- gg1 +
  facet_wrap(~ Region, ncol = 5) +
  theme(axis.text.x = element_text(angle = 30, hjust = 1))
)
```

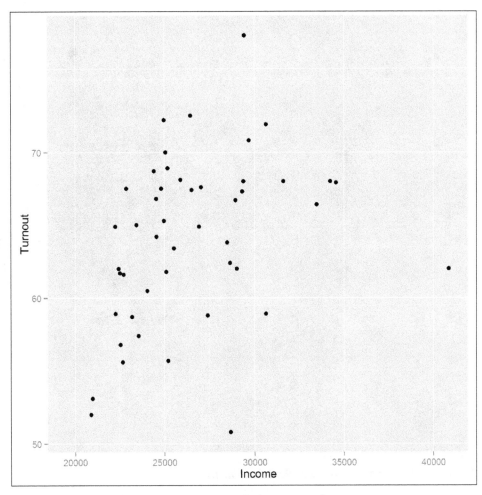

Figure 14-17. This plot is stored as a variable that is reused in Figure 14-18

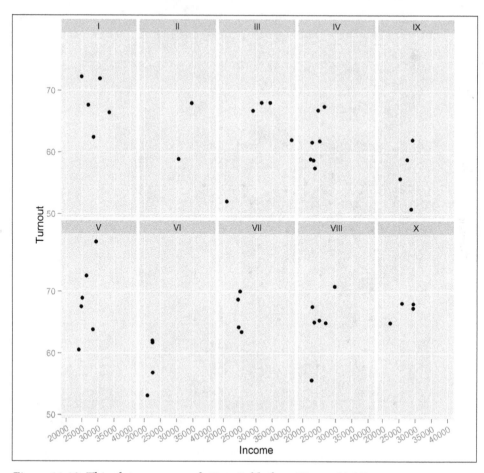

Figure 14-18. This plot reuses a ggplot2 variable from Figure 14-17

Line Plots

For exploring how a continuous variable changes over time, a line plot often provides more insight than a scatterplot, since it displays the connections between sequential values. These next examples examine a year in the life of the crab in the `crab_tag` dataset, and how deep in the North Sea it went.

In `base`, line plots are created in the same way as scatterplots, except that they take the argument `type = "l"`. To avoid any dimensional confusion[6] we plot the depth as a negative number rather than using the absolute values given in the dataset.

---

6. Insert your own Australia/upside-down joke here.

Ranges in the plot default to the ranges of the data (plus a little bit more; see the xaxs section of the ?par help page for the exact details). To get a better sense of perspective, we'll manually set the y-axis limit to run from the deepest point that the crab went in the sea up to sea level, by passing a ylim argument. Figure 14-19 displays the resulting line plot:

```
with(
  crab_tag$daylog,
  plot(Date, -Max.Depth, type = "l", ylim = c(-max(Max.Depth), 0))
)
```

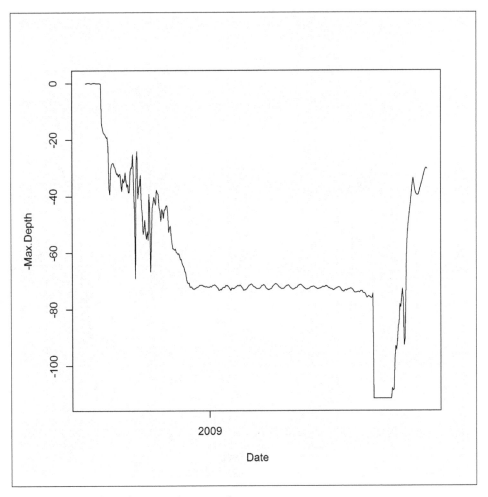

Figure 14-19. A line plot using base graphics

At the moment, this only shows half the story. The Max.Depth argument is the deepest point in the sea that the crab reached on a given day. We also need to add a line for the Min.Depth to see the shallowest point on each day. Additional lines can be drawn on an existing plot using the lines function. The equivalent for scatterplots is points. Figure 14-20 shows the additional line:

```
with(
  crab_tag$daylog,
  lines(Date, -Min.Depth, col = "blue")
)
```

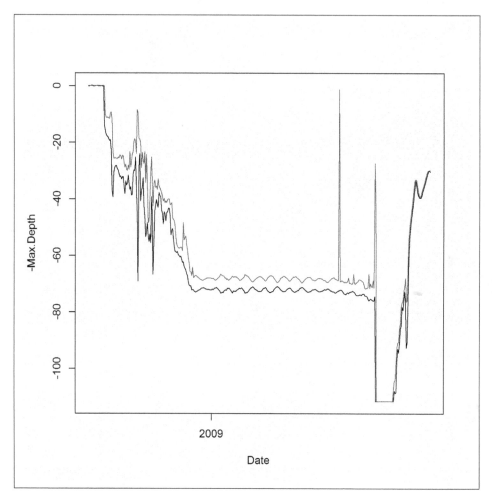

Figure 14-20. Adding a second line using base graphics

Line plots in `lattice` follow a similar pattern to `base`. They use `xyplot`, as with scatterplots, and require the same `type = "l"` argument. Specifying multiple lines is blissfully easy using the formula interface. Notice the + in the formula used to create the plot in Figure 14-21:

```
xyplot(-Min.Depth + -Max.Depth ~ Date, crab_tag$daylog, type = "l")
```

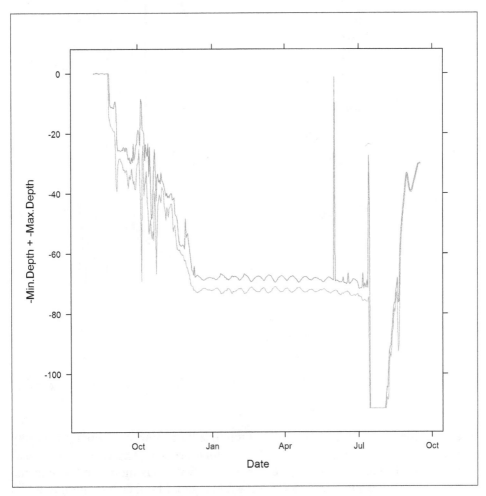

Figure 14-21. A line plot using lattice graphics

In `ggplot2`, swapping a scatterplot for a line plot is as simple as swapping `geom_plot` for `geom_line` (Figure 14-22 shows the result):

```
ggplot(crab_tag$daylog, aes(Date, -Min.Depth)) +
  geom_line()
```

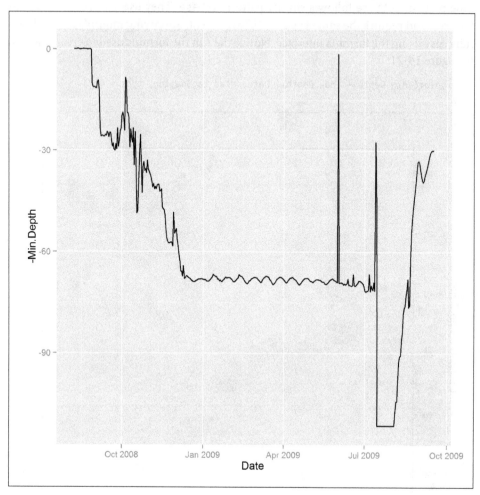

Figure 14-22. A line plot using ggplot2 graphics

There's a little complication with drawing multiple lines, however. When you specify aesthetics in the call to ggplot, you specify them for every geom. That is, they are "global" aesthetics for the plot. In this case, we want to specify the maximum depth in one line and the minimum depth in another, as shown in Figure 14-23. One solution to this is to specify the y-aesthetic inside each call to geom_line:

```
ggplot(crab_tag$daylog, aes(Date)) +
  geom_line(aes(y = -Max.Depth)) +
  geom_line(aes(y = -Min.Depth))
```

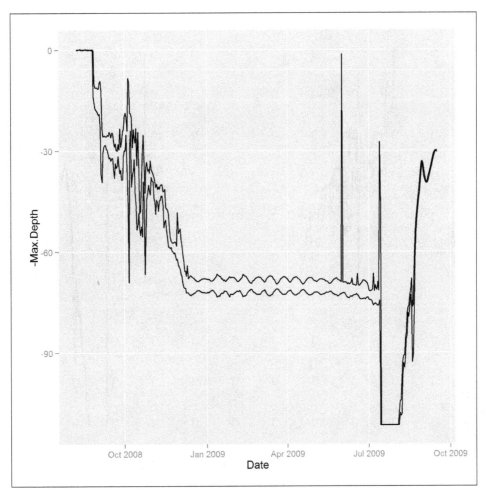

Figure 14-23. Two lines with separate geoms using ggplot2 graphics

This is a bit clunky, though, as we have to call `geom_line` twice, and actually it isn't a very idiomatic solution. The "proper" `ggplot2` way of doing things, shown in Figure 14-24, is to melt the data to long form and then group the lines:

```
library(reshape2)
crab_long <- melt(
  crab_tag$daylog,
  id.vars     = "Date",
  measure.vars = c("Min.Depth", "Max.Depth")
)
ggplot(crab_long, aes(Date, -value, group = variable)) +
  geom_line()
```

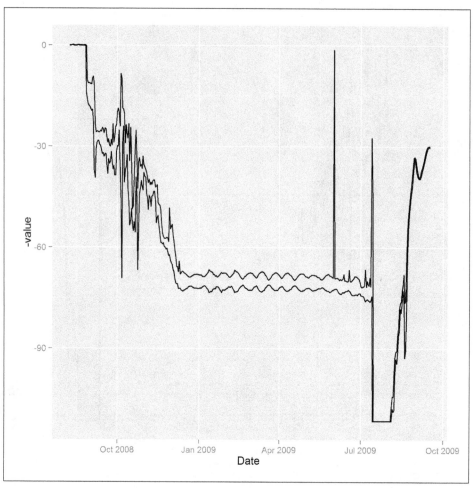

Figure 14-24. Two lines with grouping using ggplot2 graphics

In this case, where there are only two lines, there is an even better solution that doesn't require any data manipulation. geom_ribbon plots two lines, and the contents in between. For prettiness, we pass the color and fill argument to the geom, specifying the color of the lines and the bit in between. Figure 14-25 shows the result:

```
ggplot(crab_tag$daylog, aes(Date, ymin = -Min.Depth, ymax = -Max.Depth)) +
    geom_ribbon(color = "black", fill = "white")
```

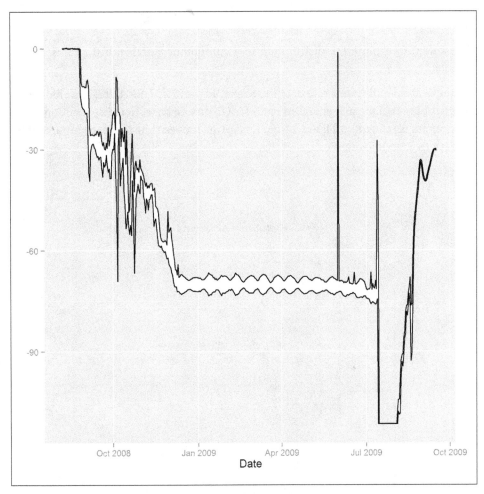

Figure 14-25. A ribbon plot using ggplot2 graphics

Whichever system you use to draw the plot, the behavior of the crab is clear. In September it lives in shallow waters for the mating season, then it spends a few months migrating into deeper territory. Through winter, spring, and summer it happily sits on the North Sea seabed (except for an odd, brief trip to the surface at the start of June—dodgy data, or a narrow escape from a fishing boat?), then it apparently falls off a cliff in mid-July, before making its way back to shallow climes for another round of rumpy-pumpy, at which point it is caught.

Histograms

If you want to explore the distribution of a continuous variable, histograms are the obvious choice.[7]

For the next examples we'll return to the obama_vs_mccain dataset, this time looking at the distribution of the percentage of votes for Obama. In base, the hist function draws a histogram, as shown in Figure 14-26. Like plot, it doesn't have a data argument, so we have to wrap it inside a call to with:

```
with(obama_vs_mccain, hist(Obama))
```

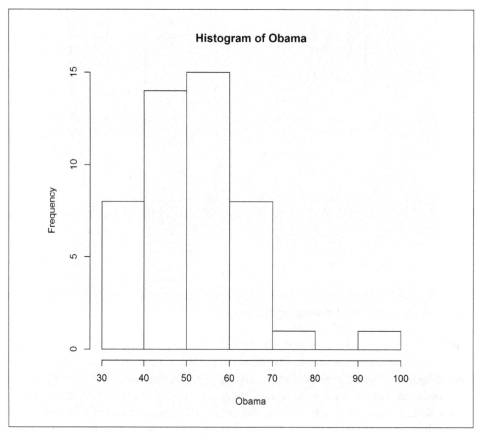

Figure 14-26. A histogram using base graphics

---

7. Dataviz purists often note that kernel density plots generally give a "better" representation of the underlying distribution. The downside is that every time you show them to a non-statistician, you have to spend 15 minutes explaining what a kernel density plot is.

The number of breaks is calculated by default by Sturges's algorithm. It is good practice to experiment with the width of bins in order to get a more complete understanding of the distribution. This can be done in a variety of ways: you can pass hist a single number to specify the number of bins, or a vector of bin edges, or the name of a different algorithm for calculating the number of bins ("scott" and "fd" are currently supported on top of the default of "sturges"), or a function that calculates one of the first two options. It's really flexible. In the following examples, the results of which are shown in Figures 14-27 to 14-31, the main argument creates a main title above the plot. It works for the plot function too:

```
with(obama_vs_mccain,
  hist(Obama, 4, main = "An exact number of bins")
)
#Fig. 14-27
```

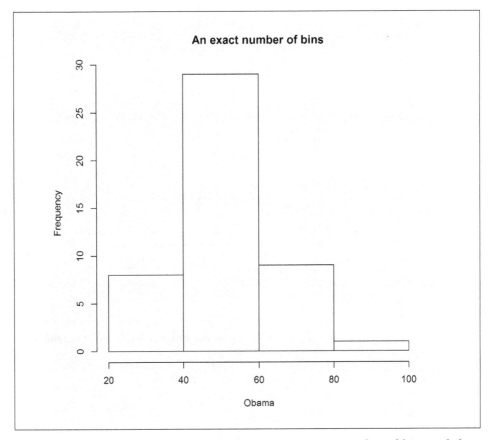

Figure 14-27. Specifying histogram breaks using an exact number of bins with base graphics

```
with(obama_vs_mccain,
  hist(Obama, seq.int(0, 100, 5), main = "A vector of bin edges")
)
#Fig. 14-28
```

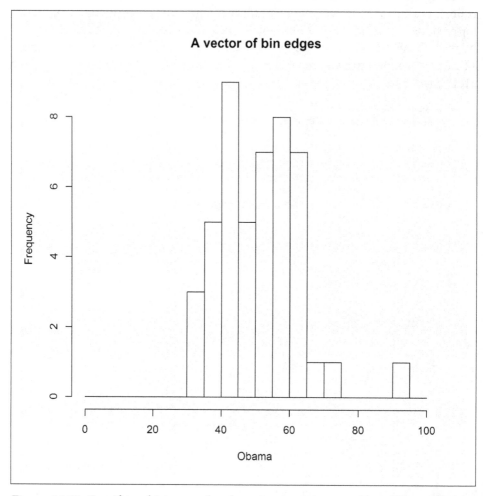

Figure 14-28. Specifying histogram breaks using an exact number of bins with base graphics

```
with(obama_vs_mccain,
    hist(Obama, "FD", main = "The name of a method")
)
#Fig. 14-29
```

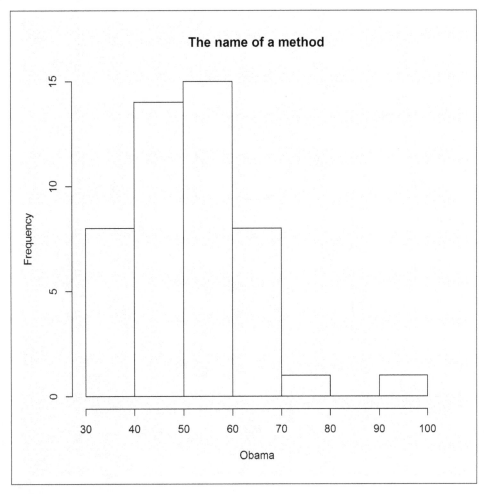

Figure 14-29. Specifying histogram breaks using the name of a method with base graphics

```
with(obama_vs_mccain,
  hist(Obama, nclass.scott, main = "A function for the number of bins")
)
#Fig. 14-30
```

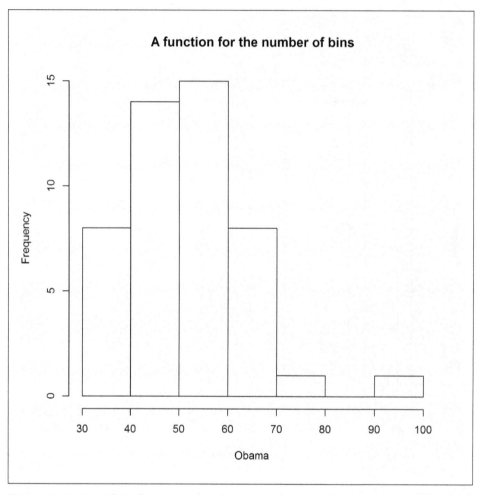

Figure 14-30. Specifying histogram breaks using a function for the number of bins with base graphics

```
binner <- function(x)
{
  seq(min(x, na.rm = TRUE), max(x, na.rm = TRUE), length.out = 50)
}
with(obama_vs_mccain,
  hist(Obama, binner, main = "A function for the bin edges")
)
#Fig. 14-31
```

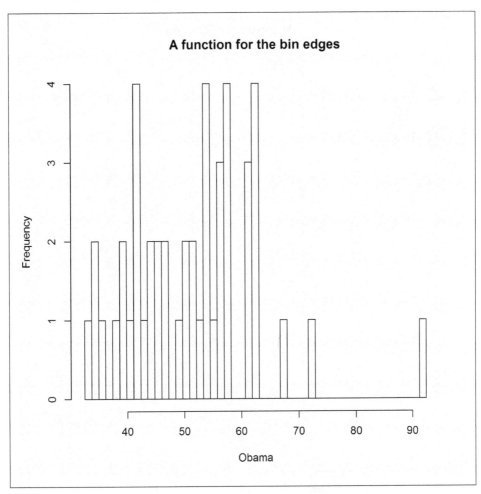

Figure 14-31. Specifying histogram breaks using a function for the bin edges with base graphics

The `freq` argument controls whether or not the histogram shows counts or probability densities in each bin. It defaults to TRUE if and only if the bins are equally spaced. Figure 14-32 shows the output:

```
with(obama_vs_mccain, hist(Obama, freq = FALSE))
```

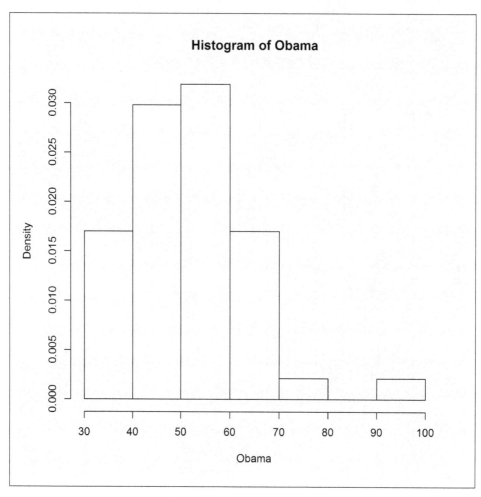

Figure 14-32. A probability density histogram using base graphics

lattice histograms behave in a similar manner to base ones, except for the usual benefits of taking a data argument, allowing easy splitting into panels, and saving plots as a variable. The breaks argument behaves in the same way as with hist. Figures 14-33 and 14-34 show lattice histograms and the specification of breaks:

```
histogram(~ Obama, obama_vs_mccain)
#Fig. 14-33

histogram(~ Obama, obama_vs_mccain, breaks = 10)
#Fig. 14-34
```

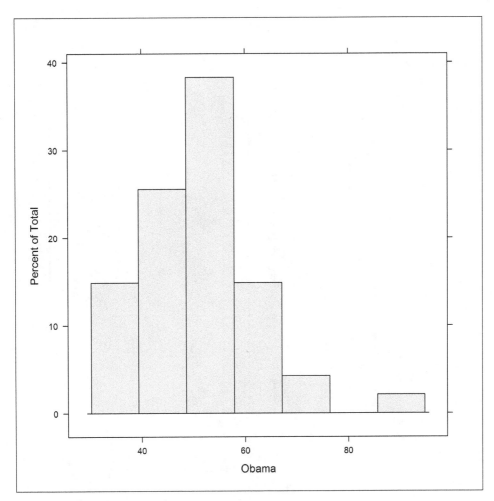

Figure 14-33. Histogram using lattice graphics

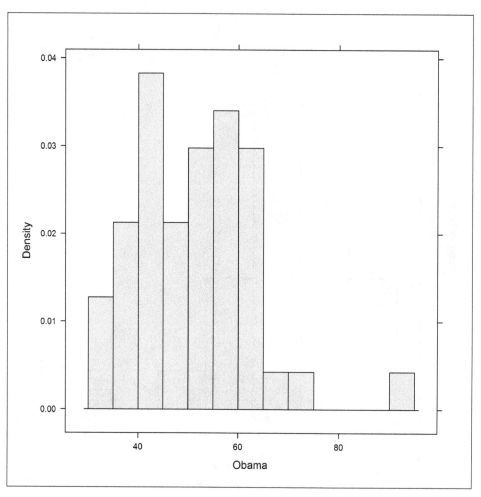

Figure 14-34. Specifying histogram breaks using lattice graphics

lattice histograms support counts, probability densities, and percentage y-axes via the type argument, which takes the string "count", "density", or "percent". Figure 14-35 shows the "percent" style:

```
histogram(~ Obama, obama_vs_mccain, type = "percent")
```

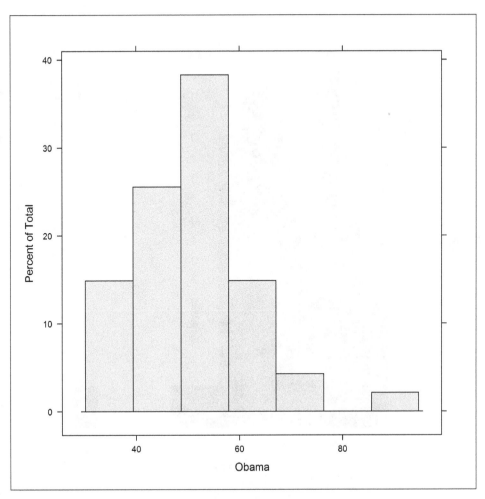

Figure 14-35. A percentage-scaled histogram using lattice graphics

ggplot2 histograms are created by adding a histogram geom. Bin specification is simple: just pass a numeric bin width to geom_histogram. The rationale is to force you to manually experiment with different numbers of bins, rather than settling for the default. Figure 14-36 shows the usage:

```
ggplot(obama_vs_mccain, aes(Obama)) +
  geom_histogram(binwidth = 5)
```

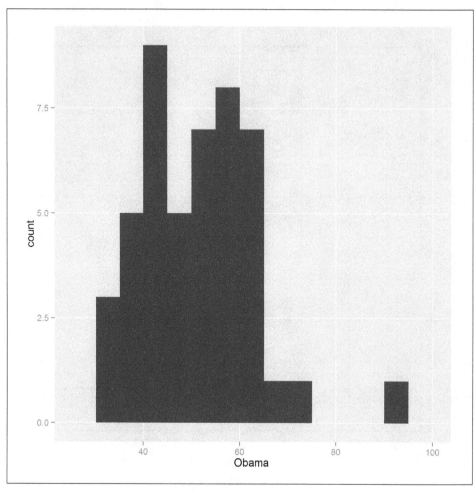

Figure 14-36. A histogram using ggplot2 graphics

You can choose between counts and densities by passing the special names `..count..` or `..density..` to the y-aesthetic. Figure 14-37 demonstrates the use of `..density..`:

```
ggplot(obama_vs_mccain, aes(Obama, ..density..)) +
  geom_histogram(binwidth = 5)
```

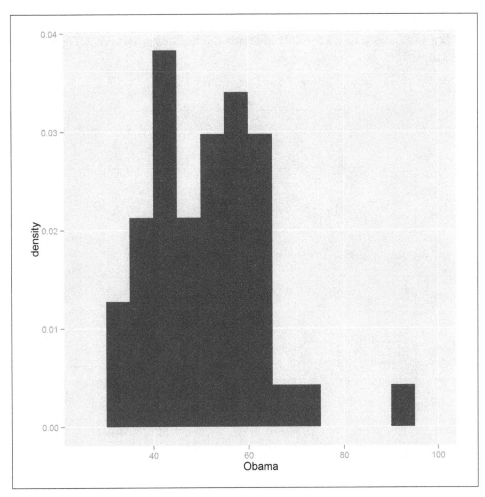

Figure 14-37. A probability density histogram using ggplot2 graphics

Box Plots

If you want to explore the distribution of lots of related variables, you could draw lots of histograms. For example, if you wanted to see the distribution of Obama votes by US region, you could use latticing/faceting to draw 10 histograms. This is just about feasible, but it doesn't scale much further. If you need a hundred histograms, the space requirements can easily overwhelm the largest monitor. Box plots (sometimes called box and whisker plots) are a more space-efficient alternative that make it easy to compare many distributions at once. You don't get as much detail as with a histogram or kernel density plot, but simple higher-or-lower and narrower-or-wider comparisons can easily be made.

The `base` function for drawing box plots is called `boxplot`; it is heavily inspired by `lattice`, insofar as it uses a formula interface and has a `data` argument. Figure 14-38 shows the usage:

```
boxplot(Obama ~ Region, data = obama_vs_mccain)
```

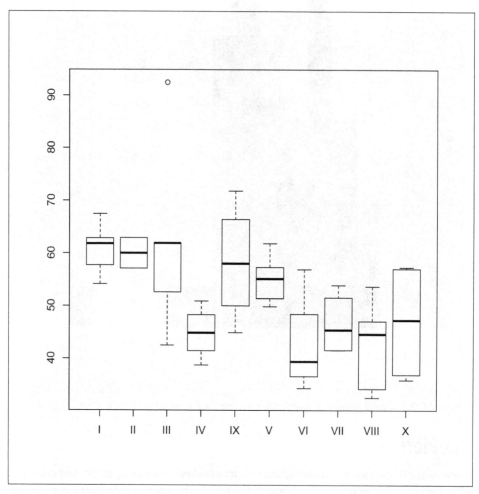

Figure 14-38. A box plot using base graphics

This type of plot is often clearer if we reorder the box plots from smallest to largest, in some sense. The `reorder` function changes the order of a factor's levels, based upon some numeric score. In Figure 14-39 we score the `Region` levels by the median Obama value for each region:

```
ovm <- within(
  obama_vs_mccain,
  Region <- reorder(Region, Obama, median)
)
boxplot(Obama ~ Region, data = ovm)
```

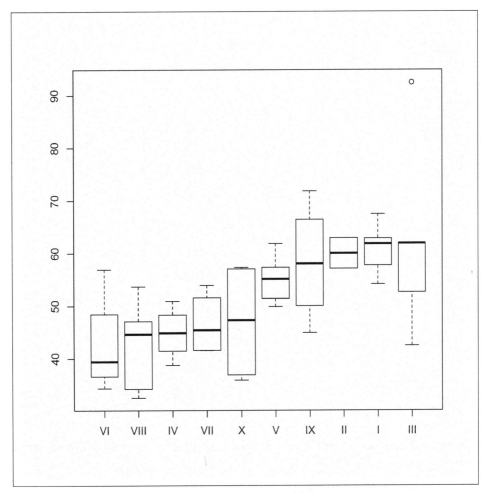

Figure 14-39. Ordering boxes using base graphics

The switch from base to lattice is very straightforward. In this simplest case, we can make a straight swap of boxplot for bwplot ("bw" is short for "b (box) and w (whisker)," in case you hadn't figured it out). Notice the similarity of Figure 14-40 to Figure 14-38:

```
bwplot(Obama ~ Region, data = ovm)
```

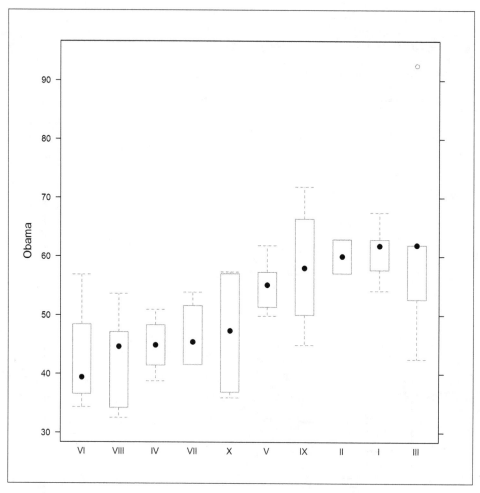

Figure 14-40. A box plot using lattice graphics

The `ggplot2` equivalent box plot, shown in Figure 14-41, just requires that we add a `geom_boxplot`:

```
ggplot(ovm, aes(Region, Obama)) +
  geom_boxplot()
```

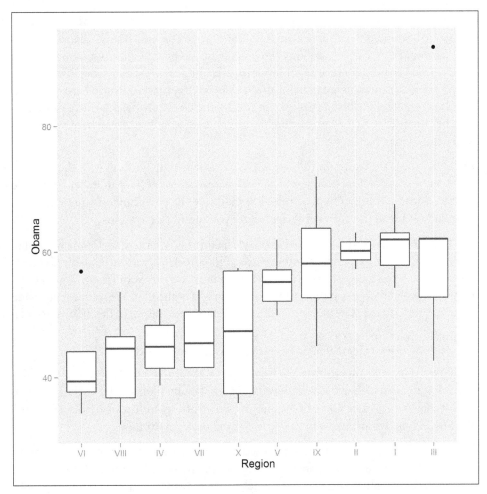

Figure 14-41. A box plot using ggplot2 graphics

Bar Charts

Bar charts (a.k.a. bar plots) are the natural way of displaying numeric variables[8] split by a categorical variable. In the next examples, we look at the distribution of religious identification across the US states. Data for Alaska and Hawaii are not included in the dataset, so we can remove those records:

```
ovm <- ovm[!(ovm$State %in% c("Alaska", "Hawaii")), ]
```

8. More specifically, they have to be counts or lengths or other numbers that can be compared to zero. For log-scaled things where the bar would extend to minus infinity, you want a dot plot instead.

In `base`, bar charts are created with the `barplot` function. As with the `plot` function, there is no argument to specify a data frame, so we need to wrap it in a call to `with`. The first argument to `barplot` contains the lengths of the bars. If that is a named vector (which it won't be if you are doing things properly and accessing data from inside a data frame), then those names are used for the bar labels. Otherwise, as we do here, you need to pass an argument called `names.arg` to specify the labels. By default the bars are vertical, but in order to make the state names readable we want horizontal bars, which can be generated with `horiz = TRUE`.

To display the state names in full, we also need to do some fiddling with the plot parameters, via the `par` function. For historical reasons, most of the parameter names are abbreviations rather than human-readable values, so the code can look quite terse. It's a good idea to read the `?par` help page before you modify a `base` plot.

The `las` parameter (short for "label axis style") controls whether labels are horizontal, vertical, parallel, or perpendicular to the axes. Plots are usually more readable if you set `las = 1`, for horizontal. The `mar` parameter is a numeric vector of length 4, giving the width of the plot margins at the bottom/left/top/right of the plot. We want a really wide lefthand side to fit the state names. Figure 14-42 shows the output of the following code:

```
par(las = 1, mar = c(3, 9, 1, 1))
with(ovm, barplot(Catholic, names.arg = State, horiz = TRUE))
```

Simple bar charts like this are fine, but more interesting are bar charts of several variables at once. We can visualize the split of religions by state by plotting the `Catholic`, `Prot estant`, `Non.religious`, and `Other` columns. For plotting multiple variables, we must place them into a matrix, one in each row (`rbind` is useful for this).

The column names of this matrix are used for the names of the bars; if there are no column names we must pass a `names.arg` like we did in the last example. By default, the bars for each variable are drawn next to each other, but since we are examining the split between the variables, a stacked bar chart is more appropriate. Passing `beside = FALSE` achieves this, as illustrated in Figure 14-43:

```
religions <- with(ovm, rbind(Catholic, Protestant, Non.religious, Other))
colnames(religions) <- ovm$State
par(las = 1, mar = c(3, 9, 1, 1))
barplot(religions, horiz = TRUE, beside = FALSE)
```

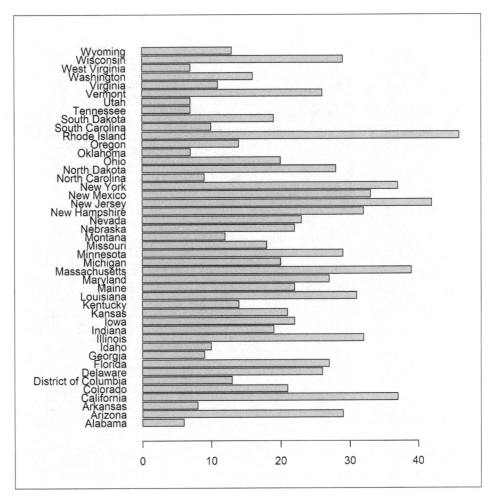

Figure 14-42. A bar chart using base graphics

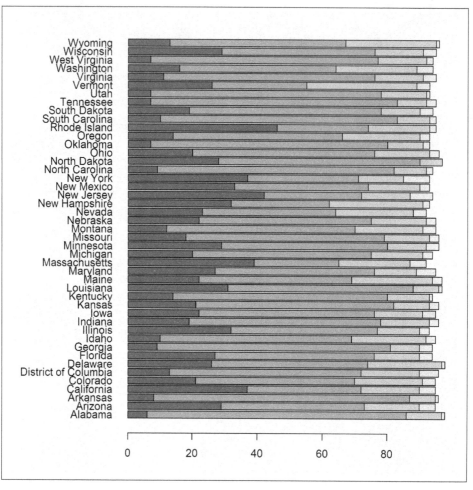

Figure 14-43. A stacked bar chart using base graphics

The lattice equivalent of barplot, shown in Figure 14-44, is barchart. The formula interface is the same as those we saw with scatterplots, yvar ~ xvar:

```
barchart(State ~ Catholic, ovm)
```

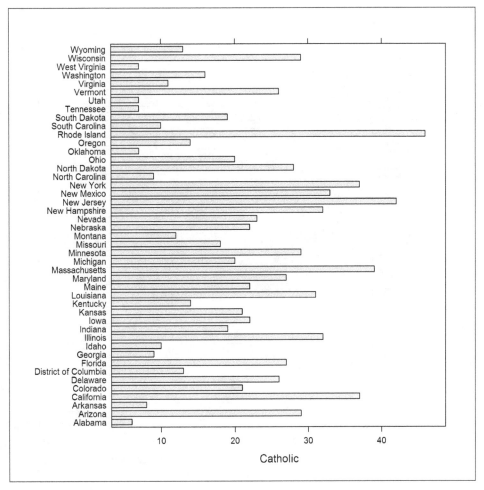

Figure 14-44. A bar chart using lattice graphics

Extending this to multiple variables just requires a tweak to the formula, and passing stack = TRUE to make a stacked plot (see Figure 14-45):

```
barchart(
    State ~ Catholic + Protestant + Non.religious + Other,
    ovm,
    stack = TRUE
)
```

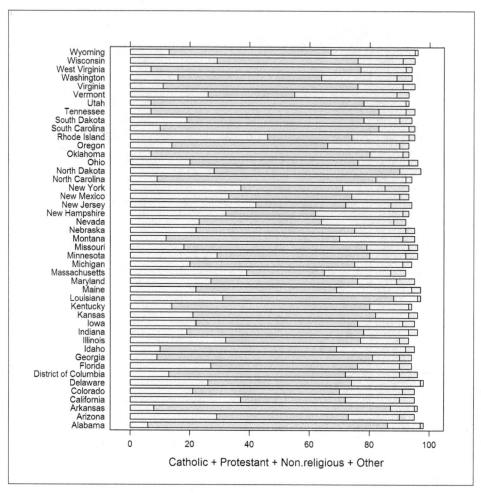

Figure 14-45. A stacked bar chart using lattice graphics

ggplot2 requires a tiny bit of work be done to the data to replicate this plot. We need the data in long form, so we must first melt the columns that we need:

```
religions_long <- melt(
  ovm,
  id.vars = "State",
  measure.vars = c("Catholic", "Protestant", "Non.religious", "Other")
)
```

Like base, gplot2 defaults to vertical bars; adding coord_flip swaps this. Finally, since we already have the lengths of each bar in the dataset (without further calculation) we must pass stat = "identity" to the geom. Bars are stacked by default, as shown in Figure 14-46:

```
ggplot(religions_long, aes(State, value, fill = variable)) +
  geom_bar(stat = "identity") +
  coord_flip()
```

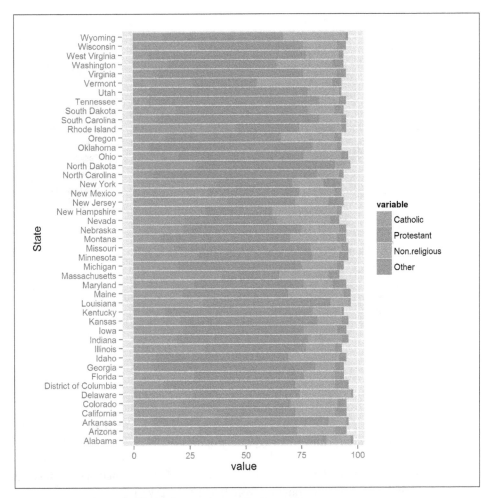

Figure 14-46. A stacked bar chart using ggplot2 graphics

To avoid the bars being stacked, we would have to pass the argument `position =`
`"dodge"` to `geom_bar`. Figure 14-47 shows this:

```
ggplot(religions_long, aes(State, value, fill = variable)) +
  geom_bar(stat = "identity", position = "dodge") +
  coord_flip()
```

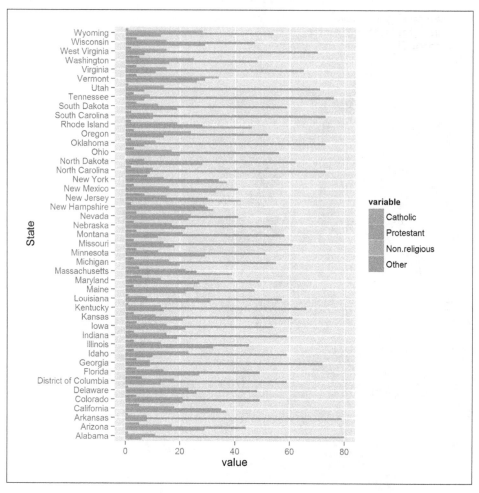

Figure 14-47. A dodged bar chart using ggplot2 graphics

The other possibility for that argument is `position = "fill"`, which creates stacked bars that are all the same height, ranging from 0 to 100%. Try it!

Other Plotting Packages and Systems

There are many packages that contain plotting capabilities based on one or more of the three systems. For example, the `vcd` package has lots of plots for visualizing categorical data, such as mosaic plots and association plots. `plotrix` has loads of extra plot types, and there are specialist plots scattered in many other packages.

`latticeExtra` and `GGally` extend the `lattice` and `ggplot2` packages, and `grid` provides access to the underlying framework that supports both these systems.

You may have noticed that all the plots covered so far are static. There have in fact been a number of attempts to provide dynamic and interactive plots.[9] There is no perfect solution yet, but there are many interesting and worthy packages that attempt this.

gridSVG lets you write grid-based plots (lattice or ggplot2) to SVG files. These can be made interactive, but it requires some knowledge of JavaScript. playwith allows pointing and clicking to interact with base or lattice plots. iplots provides a whole extra system of plots with even more interactivity. It isn't easily extensible, but the common plots types are there, and you can explore the data very quickly via mouse interaction. googleVis provides an R wrapper around Google Chart Tools, creating plots that can be displayed in a browser. rggobi provides an interface to GGobi (for visualizing high-dimensional data), and rgl provides an interface to OpenGL for interactive 3D plots. The animation package lets you make animated GIFs or SWF animations.

The rCharts package provides wrappers to half a dozen JavaScript plotting libraries using a lattice syntax. It isn't yet available via CRAN, so you'll need to install it from GitHub:

```
library(devtools)
install_github("rCharts", "ramnathv")
```

Summary

- There are loads of summary statistics that can be calculated.
- R has three plotting systems: base, lattice, and ggplot2.
- All the common plot types are supported in every system.
- There is some support in R for dynamic and interactive plotting.

Test Your Knowledge: Quiz

Question 14-1
　　What is the difference between the min and pmin functions?

Question 14-2
　　How would you change the shape of the points in a base plot?

Question 14-3
　　How do you specify the *x* and *y* variables in a lattice plot?

---

9. Dynamic means animation; interactive means point and click to change them.

Question 14-4

What is a `ggplot2` *aesthetic*?

Question 14-5

Name as many plot types as you can think of for exploring the distribution of a continuous variable.

Test Your Knowledge: Exercises

Exercise 14-1

1. In the `obama_vs_mccain` dataset, find the (Pearson) correlation between the percentage of unemployed people within the state and the percentage of people that voted for Obama. [5]

2. Draw a scatterplot of the two variables, using a graphics system of your choice. (For bonus points, use all three systems.) [10] for one plot, [30] for all three

Exercise 14-2

In the `alpe_d_huez2` dataset, plot the distributions of fastest times, split by whether or not the rider (allegedly) used drugs. Display this using a) histograms and b) box plots. [10]

Exercise 14-3

The `gonorrhoea` dataset contains gonorrhoea infection rates in the US by year, age, ethnicity, and gender. Explore how infection rates change with age. Is there a time trend? Do ethnicity and gender affect the infection rates? [30]

## CHAPTER 15

Distributions and Modeling

Using summary statistics and plots to understand data is great, but they have their limitations. Statistics don't give you the shape of data, and plots aren't scalable to many variables (with more than five or six, things start to get confusing),[1] nor are they scalable in number (since you have to physically look at each one). Neither statistics nor plots are very good at giving you predictions from the data.

This is where models come in: if you understand enough about the structure of the data to be able to run a suitable model, then you can pass quantitative judgments about the data and make predictions.

There are lots of different statistical models, with more being invented as fast as university statistics departments can think of them. In order to avoid turning into a statistics course, this chapter is just going to deal with some really simple regression models. If you want to learn some stats, I recommend *The R Book* or *Discovering Statistics Using R*, both of which explain statistical concepts in glorious slow motion.

Before we get to running any models, we need a bit of background on generating random numbers, different kinds of distributions, and formulae.

Chapter Goals

After reading this chapter, you should:

- Be able to generate random numbers from many distributions
- Be able to find quantiles and inverse quantiles from those distributions

---

1. If you've just started casually doodling a biplot, then well done, have some geek points. Plots of lots of variables are possible if you use a dimension-reduction technique like principal component analysis or factor analysis to give you fewer variables in practice.

- Know how to write a model formula
- Understand how to run, update, and plot a linear regression

Random Numbers

Random numbers are critical to many analyses, and consequently R has a wide variety of functions for sampling from different distributions.

The sample Function

We've seen the `sample` function a few times already (it was first introduced in Chapter 3). It's an important workhorse function for generating random values, and its behavior has a few quirks, so it's worth getting to know it more thoroughly. If you just pass it a number, *n*, it will return a permutation of the natural numbers from 1 to *n*:

```
sample(7)
## [1] 1 2 5 7 4 6 3
```

If you give it a second value, it will return that many random numbers between 1 and *n*:

```
sample(7, 5)
## [1] 7 2 3 1 5
```

Notice that all those random numbers are different. By default, `sample` samples without replacement. That is, each value can only appear once. To allow sampling with replacement, pass `replace = TRUE`:

```
sample(7, 10, replace = TRUE)
##  [1] 4 6 1 7 5 3 6 7 4 2
```

Of course, returning natural numbers isn't that interesting most of the time, but `sample` is flexible enough to let us sample from any vector that we like. Most commonly, we might want to pass it a character vector:

```
sample(colors(), 5) #a great way to pick the color scheme for your house
## [1] "grey53"      "deepskyblue2" "gray94"       "maroon2"
## [5] "gray18"
```

If we were feeling more adventurous, we could pass it some dates:

```
sample(.leap.seconds, 4)
## [1] "2012-07-01 01:00:00 BST" "1994-07-01 01:00:00 BST"
## [3] "1981-07-01 01:00:00 BST" "1990-01-01 00:00:00 GMT"
```

We can also weight the probability of each input value being returned by passing a `prob` argument. In the next example, we use R to randomly decide which month to go on holiday, then fudge it to increase our chances of a summer break:

```
weights <- c(1, 1, 2, 3, 5, 8, 13, 21, 8, 3, 1, 1)
sample(month.abb, 1, prob = weights)
```

```
## [1] "Jul"
```

Sampling from Distributions

Oftentimes, we want to generate random numbers from a probability distribution. R has functions available for sampling from almost one hundred distributions, mixtures, and copulas across the various packages. The `?Distribution` help page documents the facilities available in base R, and the CRAN Task View on distributions (*http://cran.r-project.org/web/views/Distributions.html*) gives extensive directions about which package to look in if you want something more esoteric.

Most of the random number generation functions have the name `r<distn>`. For example, we've already seen `runif`, which generates uniform random numbers, and `rnorm`, which generates normally distributed random numbers. The first parameter for each of these functions is the number of random numbers to generate, and further parameters affect the shape of the distribution. For example, `runif` allows you to set the lower and upper bounds of the distribution:

```
runif(5)         #5 uniform random numbers between 0 and 1
runif(5, 1,10)   #5 uniform random numbers between 1 and 10
rnorm(5)         #5 normal random numbers with mean 0 and std dev 1
rnorm(5, 3, 7)   #5 normal random numbers with mean 3 and std dev 7
```

Random numbers generated by R are, like with any other software, actually pseudorandom. That is, they are generated by an algorithm rather than a genuinely random process. R supports several algorithms out of the box, as described on the ? RNG page, and more specialist algorithms (for parallel number generation, for example) in other packages. The CRAN Task View on distributions mentioned above also describes where to find more algorithms. You can see which algorithms are used for uniform and normal random number generation with the `RNGkind` function (sampling from other distributions typically uses some function of uniform or normal random numbers):

```
RNGkind()
```

```
## [1] "Mersenne-Twister" "Inversion"
```

Random number generators require a starting point to generate numbers from, known as a "seed." By setting the seed to a particular value, you can guarantee that the same random numbers will be generated each time you run the same piece of code. For example, this book fixes the seed so that the examples are the same each time the book is built. You can set the seed with the `set.seed` function. It takes a positive integer as an input. Which integer you choose doesn't matter; different seeds just give different random numbers:

```
set.seed(1)
runif(5)

## [1] 0.2655 0.3721 0.5729 0.9082 0.2017

set.seed(1)
runif(5)

## [1] 0.2655 0.3721 0.5729 0.9082 0.2017

set.seed(1)
runif(5)

## [1] 0.2655 0.3721 0.5729 0.9082 0.2017
```

You can also specify different generation algorithms, though this is very advanced usage, so don't do that unless you know what you are doing.

Distributions

As well as a function for generating random numbers, most distributions also have functions for calculating their probability density function (PDF), cumulative density function (CDF), and inverse CDF.

Like the RNG functions with names beginning with r, these functions have names beginning with d, p, and q, respectively. For example, the normal distribution has functions `dnorm`, `pnorm`, and `qnorm`. These functions are perhaps best demonstrated visually, and are shown in Figure 15-1. (The code is omitted, on account of being rather fiddly.)

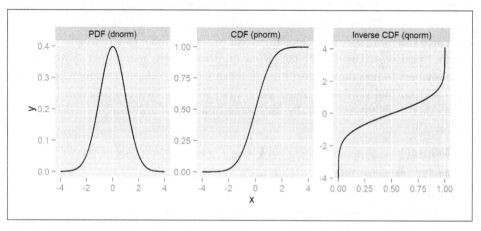

Figure 15-1. PDF, CDF, and inverse CDF of a normal distribution

Formulae

We've already seen the use of formulae in `lattice` plots and `ggplot2` facets. Many statistical models use formulae in a similar way, in order to specify the structure of the variables in the model. The exact meaning of the formula depends upon the model function that it gets passed to (in the same way that a formula in a `lattice` plot means something different to a formula in a `ggplot2` facet), but there is a common pattern that most models satisfy: the lefthand side specifies the response variable, the righthand side specifies independent variables, and the two are separated by a tilde. Returning to the `gonorrhoea` dataset as an example, we have:

```
Rate ~ Year + Age.Group + Ethnicity + Gender
```

Here, `Rate` is the response, and we have chosen to include four independent variables (year/age group/ethnicity/gender). For models that can include an intercept (that's more or less all regression models), a formula like this will implicitly include that intercept. If we passed it into a linear regression model, it would mean:

$$Rate = \alpha_0 + \alpha_1 * Year + \alpha_2 * Age.Group + \alpha_3 * Ethnicity + \alpha_4 * Gender + \epsilon$$

where each α_i is a constant to be determined by the model, and ϵ is a normally distributed error term.

If we didn't want the intercept term, we could add a zero to the righthand side to suppress it:

```
Rate ~ 0 + Year + Age.Group + Ethnicity + Gender
```

The updated equation given by this formula is:

$$Rate = \alpha_1 * Year + \alpha_2 * Age.Group + \alpha_3 * Ethnicity + \alpha_4 * Gender + \epsilon$$

Each of these formulae includes only the individual independent variables, without any interactions between them. To include interactions, we can replace the plusses with asterisks:

```
Rate ~ Year * Age.Group * Ethnicity * Gender
```

This adds in all possible two-way interactions (year and age group, year and ethnicity, etc.), and three-way interactions (year and age group and ethnicity, etc.), all the way up to an every-way interaction between all independent variables (year and age group and ethnicity and gender). This is very often overkill, since a every-way interaction is usually meaningless.

There are two ways of restricting the level of interaction. First, you can add individual interactions using colons. In the following example, Year:Ethnicity is a two-way interaction between those terms, and Year:Ethnicity:Gender is a three-way interaction:

```
Rate ~ Year + Ethnicity + Gender + Year:Ethnicity + Year:Ethnicity:Gender
```

This fine-grained approach can become cumbersome to type if you have more than three or four variables, though, so an alternate syntax lets you include all interactions up to a certain level using carets (^). The next example includes the year, ethnicity, and gender, and the three two-way interactions:

```
Rate ~ (Year + Ethnicity + Gender) ^ 2
```

You can also include modified versions of variables. For example, a surprising number of environmental processes generate lognormally distributed variables, which you may want to include in a linear regression as log(var). Terms like this can be included directly, but you may have spotted a problem with including var ^ 2. That syntax is reserved for interactions, so if you want to include powers of things, wrap them in I():

```
Rate ~ I(Year ^ 2)  #year squared, not an interaction
```

A First Model: Linear Regressions

Ordinary least squares linear regressions are the simplest in an extensive family of regression models. The function for calculating them is concisely if not clearly named: lm, short for "linear model." It accepts a formula of the type we've just discussed, and a data frame that contains the variables to model. Let's take a look at the gonorrhoea dataset. For simplicity, we'll ignore interactions:

```
model1 <- lm(Rate ~ Year + Age.Group + Ethnicity + Gender, gonorrhoea)
```

If we print the model variable, it lists the coefficients for each input variable (the α_i values). If you look closely, you'll notice that for both of the categorical variables that we put into the model (age group and ethnicity), one category has no coefficient. For example, the 0 to 4 age group is missing, and the American Indians & Alaskan Natives ethnicity is also missing.

These "missing" categories are included in the intercept. In the following output, the intercept value of 5,540 people infected with gonorrhoea per 100,000 people applies to 0- to 4-year-old female American Indians and Alaskan Natives in the year 0. To predict infection rates in the years up to 2013, we would add 2013 times the coefficient for Year, -2.77. To predict the effect on 25- to 29-year-olds of the same ethnicity, we would add the coefficient for that age group, 291:

```
model1
```

```
## 
## Call:
## lm(formula = Rate ~ Year + Age.Group + Ethnicity + Gender, data = gonorrhoea)
## 
## Coefficients:
##                        (Intercept)                                Year
##                           5540.496                              -2.770
##                     Age.Group5 to 9                   Age.Group10 to 14
##                             -0.614                              15.268
##                   Age.Group15 to 19                   Age.Group20 to 24
##                            415.698                             546.820
##                   Age.Group25 to 29                   Age.Group30 to 34
##                            291.098                             155.872
##                   Age.Group35 to 39                   Age.Group40 to 44
##                             84.612                              49.506
##                   Age.Group45 to 54                   Age.Group55 to 64
##                             27.364                               8.684
##                 Age.Group65 or more   EthnicityAsians & Pacific Islanders
##                              1.178                             -82.923
##                   EthnicityHispanics        EthnicityNon-Hispanic Blacks
##                            -49.000                             376.204
##        EthnicityNon-Hispanic Whites                          GenderMale
##                            -68.263                             -17.892
```

The "0- to 4-year-old female American Indians and Alaskan Natives" group was chosen because it consists of the first level of each of the factor variables. We can see those factor levels by looping over the dataset and calling levels:

```
lapply(Filter(is.factor, gonorrhoea), levels)

## $Age.Group
##  [1] "0 to 4"     "5 to 9"     "10 to 14"   "15 to 19"   "20 to 24"
##  [6] "25 to 29"   "30 to 34"   "35 to 39"   "40 to 44"   "45 to 54"
## [11] "55 to 64"   "65 or more"
## 
## $Ethnicity
## [1] "American Indians & Alaskan Natives"
## [2] "Asians & Pacific Islanders"
## [3] "Hispanics"
## [4] "Non-Hispanic Blacks"
## [5] "Non-Hispanic Whites"
## 
## $Gender
## [1] "Female" "Male"
```

As well as knowing the size of the effect of each input variable, we usually want to know which variables were significant. The summary function is overloaded to work with lm to do just that. The most exciting bit of the output from summary is the coefficients table. The Estimate column shows the coefficients that we've seen already, and the fourth column, Pr(>|t|), shows the p-values. The fifth column gives a star rating: where the p-value is less than 0.05 a variable gets one star, less than 0.01 is two stars, and so on. This makes it easy to quickly see which variables had a significant effect:

```
summary(model1)

##
## Call:
## lm(formula = Rate ~ Year + Age.Group + Ethnicity + Gender, data = gonorrhoea)
##
## Residuals:
##     Min     1Q Median     3Q    Max
## -376.7 -130.6   37.1   90.7 1467.1
##
## Coefficients:
##                                     Estimate Std. Error t value Pr(>|t|)
## (Intercept)                         5540.496  14866.406    0.37   0.7095
## Year                                  -2.770      7.400   -0.37   0.7083
## Age.Group5 to 9                       -0.614     51.268   -0.01   0.9904
## Age.Group10 to 14                     15.268     51.268    0.30   0.7660
## Age.Group15 to 19                    415.698     51.268    8.11  3.0e-15
## Age.Group20 to 24                    546.820     51.268   10.67  < 2e-16
## Age.Group25 to 29                    291.098     51.268    5.68  2.2e-08
## Age.Group30 to 34                    155.872     51.268    3.04   0.0025
## Age.Group35 to 39                     84.612     51.268    1.65   0.0994
## Age.Group40 to 44                     49.506     51.268    0.97   0.3346
## Age.Group45 to 54                     27.364     51.268    0.53   0.5937
## Age.Group55 to 64                      8.684     51.268    0.17   0.8656
## Age.Group65 or more                    1.178     51.268    0.02   0.9817
## EthnicityAsians & Pacific Islanders  -82.923     33.093   -2.51   0.0125
## EthnicityHispanics                   -49.000     33.093   -1.48   0.1392
## EthnicityNon-Hispanic Blacks         376.204     33.093   11.37  < 2e-16
## EthnicityNon-Hispanic Whites         -68.263     33.093   -2.06   0.0396
## GenderMale                           -17.892     20.930   -0.85   0.3930
##
## (Intercept)
## Year
## Age.Group5 to 9
## Age.Group10 to 14
## Age.Group15 to 19                   ***
## Age.Group20 to 24                   ***
## Age.Group25 to 29                   ***
## Age.Group30 to 34                   **
## Age.Group35 to 39                   .
## Age.Group40 to 44
## Age.Group45 to 54
## Age.Group55 to 64
## Age.Group65 or more
## EthnicityAsians & Pacific Islanders *
## EthnicityHispanics
## EthnicityNon-Hispanic Blacks        ***
## EthnicityNon-Hispanic Whites        *
## GenderMale
## ---
```

```
## Signif. codes:  0 '***' 0.001 '**' 0.01 '*' 0.05 '.' 0.1 ' ' 1
##
## Residual standard error: 256 on 582 degrees of freedom
## Multiple R-squared:  0.491,   Adjusted R-squared:  0.476
## F-statistic: 33.1 on 17 and 582 DF,  p-value: <2e-16
```

Comparing and Updating Models

Rather than just accepting the first model that we think of, we often want to find a "best" model, or a small set of models that provide some insight.

This section demonstrates some metrics for measuring the quality of a model, such as p-values and log-likelihood measures. By using these metrics to compare models, you can automatically keep updating your model until you get to a "best" one.

Unfortunately, automatic updating of models like this ("stepwise regression") is a poor method for choosing a model, and it gets worse as you increase the number of input variables.

Better methods for model selection, like model training or model averaging, are beyond the scope of this book. The CrossValidated statistics Q&A site (*http://bit.ly/1a5q6IQ*) has a good list of possibilities.

 It is often a mistake to try to find a single "best" model. A good model is one that gives you insight into your problem—there may be several that do this.

In order to sensibly choose a model for our dataset, we need to understand a little about the things that affect gonorrhoea infection rates. Gonorrhoea is primarily sexually transmitted (with some transmission from mothers to babies during childbirth), so the big drivers are related to sexual culture: how much unprotected sex people have, and with how many partners.

The p-value for Year is 0.71, meaning not even close to significant. Over such a short time series (five years of data), I'd be surprised if there were any changes in sexual culture big enough to have an important effect on infection rates, so let's see what happens when we remove it.

Rather than having to completely respecify the model, we can update it using the update function. This accepts a model and a formula. We are just updating the righthand side of the formula, so the lefthand side stays blank. In the next example, . means "the terms that were already in the formula," and - (minus) means "remove this next term":

```
model2 <- update(model1, ~ . - Year)
summary(model2)
```

```
## 
## Call:
## lm(formula = Rate ~ Age.Group + Ethnicity + Gender, data = gonorrhoea)
## 
## Residuals:
##     Min    1Q Median     3Q    Max
## -377.6 -128.4   34.6   92.2 1472.6
## 
## Coefficients:
##                                   Estimate Std. Error t value Pr(>|t|)
## (Intercept)                        -25.103     43.116   -0.58   0.5606
## Age.Group5 to 9                     -0.614     51.230   -0.01   0.9904
## Age.Group10 to 14                   15.268     51.230    0.30   0.7658
## Age.Group15 to 19                  415.698     51.230    8.11  2.9e-15
## Age.Group20 to 24                  546.820     51.230   10.67  < 2e-16
## Age.Group25 to 29                  291.098     51.230    5.68  2.1e-08
## Age.Group30 to 34                  155.872     51.230    3.04   0.0025
## Age.Group35 to 39                   84.612     51.230    1.65   0.0992
## Age.Group40 to 44                   49.506     51.230    0.97   0.3343
## Age.Group45 to 54                   27.364     51.230    0.53   0.5934
## Age.Group55 to 64                    8.684     51.230    0.17   0.8655
## Age.Group65 or more                  1.178     51.230    0.02   0.9817
## EthnicityAsians & Pacific Islanders -82.923    33.069   -2.51   0.0124
## EthnicityHispanics                 -49.000     33.069   -1.48   0.1389
## EthnicityNon-Hispanic Blacks       376.204     33.069   11.38  < 2e-16
## EthnicityNon-Hispanic Whites       -68.263     33.069   -2.06   0.0394
## GenderMale                         -17.892     20.915   -0.86   0.3926
## 
## (Intercept)
## Age.Group5 to 9
## Age.Group10 to 14
## Age.Group15 to 19                         ***
## Age.Group20 to 24                         ***
## Age.Group25 to 29                         ***
## Age.Group30 to 34                         **
## Age.Group35 to 39                         .
## Age.Group40 to 44
## Age.Group45 to 54
## Age.Group55 to 64
## Age.Group65 or more
## EthnicityAsians & Pacific Islanders *
## EthnicityHispanics
## EthnicityNon-Hispanic Blacks              ***
## EthnicityNon-Hispanic Whites              *
## GenderMale
## ---
## Signif. codes:  0 '***' 0.001 '**' 0.01 '*' 0.05 '.' 0.1 ' ' 1
## 
## Residual standard error: 256 on 583 degrees of freedom
## Multiple R-squared:  0.491,  Adjusted R-squared:  0.477
## F-statistic: 35.2 on 16 and 583 DF,  p-value: <2e-16
```

The anova function computes ANalysis Of VAriance tables for models, letting you see if your simplified model is significantly different from the fuller model:

```
anova(model1, model2)

## Analysis of Variance Table
##
## Model 1: Rate ~ Year + Age.Group + Ethnicity + Gender
## Model 2: Rate ~ Age.Group + Ethnicity + Gender
##   Res.Df      RSS Df Sum of Sq    F Pr(>F)
## 1    582 38243062
## 2    583 38252272 -1     -9210 0.14  0.71
```

The p-value in the righthand column is 0.71, so removing the year term didn't significantly affect the model's fit to the data.

The Akaike and Bayesian information criteria provide alternate methods of comparing models, via the AIC and BIC functions. They use log-likelihood values, which tell you how well the models fit the data, and penalize them depending upon how many terms there are in the model (so simpler models are better than complex models). Smaller numbers roughly correspond to "better" models:

```
AIC(model1, model2)

##        df  AIC
## model1 19 8378
## model2 18 8376

BIC(model1, model2)

##        df  BIC
## model1 19 8462
## model2 18 8456
```

You can see the effects of these functions better if we create a silly model. Let's remove age group, which appears to be a powerful predictor of gonorrhoea infection rates (as it should be, since children and the elderly have much less sex than young adults):

```
silly_model <- update(model1, ~ . - Age.Group)
anova(model1, silly_model)

## Analysis of Variance Table
##
## Model 1: Rate ~ Year + Age.Group + Ethnicity + Gender
## Model 2: Rate ~ Year + Ethnicity + Gender
##   Res.Df      RSS Df Sum of Sq    F Pr(>F)
## 1    582 38243062
## 2    593 57212506 -11  -1.9e+07 26.2 <2e-16 ***
## ---
## Signif. codes:  0 '***' 0.001 '**' 0.01 '*' 0.05 '.' 0.1 ' ' 1
```

```
AIC(model1, silly_model)

##             df  AIC
## model1     19 8378
## silly_model 8 8598

BIC(model1, silly_model)

##             df  BIC
## model1     19 8462
## silly_model 8 8633
```

In the silly model, anova notes a significant difference between the models, and both the AIC and the BIC have gone up by a lot.

Returning to our quest for a non-silly model, notice that gender is nonsignificant (p = 0.39). If you did Exercise 14-3 (you did, right?), then this might be interesting to you because it looked like infection rates were higher in women from a plot. Hold that thought until the exercises at the end of this chapter. For now, let's trust the p-value and remove the gender term from the model:

```
model3 <- update(model2, ~ . - Gender)
summary(model3)

##
## Call:
## lm(formula = Rate ~ Age.Group + Ethnicity, data = gonorrhoea)
##
## Residuals:
##    Min     1Q Median     3Q    Max
## -380.1 -136.1   35.8   87.4 1481.5
##
## Coefficients:
##                                    Estimate Std. Error t value Pr(>|t|)
## (Intercept)                         -34.050     41.820   -0.81   0.4159
## Age.Group5 to 9                      -0.614     51.218   -0.01   0.9904
## Age.Group10 to 14                    15.268     51.218    0.30   0.7657
## Age.Group15 to 19                   415.698     51.218    8.12  2.9e-15
## Age.Group20 to 24                   546.820     51.218   10.68  < 2e-16
## Age.Group25 to 29                   291.098     51.218    5.68  2.1e-08
## Age.Group30 to 34                   155.872     51.218    3.04   0.0024
## Age.Group35 to 39                    84.612     51.218    1.65   0.0991
## Age.Group40 to 44                    49.506     51.218    0.97   0.3342
## Age.Group45 to 54                    27.364     51.218    0.53   0.5934
## Age.Group55 to 64                     8.684     51.218    0.17   0.8654
## Age.Group65 or more                   1.178     51.218    0.02   0.9817
## EthnicityAsians & Pacific Islanders -82.923     33.061   -2.51   0.0124
## EthnicityHispanics                  -49.000     33.061   -1.48   0.1389
## EthnicityNon-Hispanic Blacks        376.204     33.061   11.38  < 2e-16
## EthnicityNon-Hispanic Whites        -68.263     33.061   -2.06   0.0394
##
## (Intercept)
## Age.Group5 to 9
```

```
## Age.Group10 to 14
## Age.Group15 to 19                        ***
## Age.Group20 to 24                        ***
## Age.Group25 to 29                        ***
## Age.Group30 to 34                        **
## Age.Group35 to 39                        .
## Age.Group40 to 44
## Age.Group45 to 54
## Age.Group55 to 64
## Age.Group65 or more
## EthnicityAsians & Pacific Islanders *
## EthnicityHispanics
## EthnicityNon-Hispanic Blacks             ***
## EthnicityNon-Hispanic Whites              *
## ---
## Signif. codes:  0 '***' 0.001 '**' 0.01 '*' 0.05 '.' 0.1 ' ' 1
##
## Residual standard error: 256 on 584 degrees of freedom
## Multiple R-squared:  0.491,  Adjusted R-squared:  0.477
## F-statistic: 37.5 on 15 and 584 DF,  p-value: <2e-16
```

Finally, the intercept term looks to be nonsignificant. This is because the default group is 0- to 4-year-old American Indians and Alaskan Natives, and they don't have much gonorrhoea.

We can set a different default with the `relevel` function. As a non-Hispanic white person in the 30- to 34-year-old group,[2] I'm going to arbitrarily set those as the defaults. In this next example, notice that we can use the `update` function to update the data frame as well as the formula:

```
g2 <- within(
  gonorrhoea,
  {
    Age.Group <- relevel(Age.Group, "30 to 34")
    Ethnicity <- relevel(Ethnicity, "Non-Hispanic Whites")
  }
)
model4 <- update(model3, data = g2)
summary(model4)

##
## Call:
## lm(formula = Rate ~ Age.Group + Ethnicity, data = g2)
##
## Residuals:
##    Min     1Q Median     3Q    Max
## -380.1 -136.1   35.8   87.4 1481.5
##
## Coefficients:
```

---

2. Please remind me to change this for the second edition!

```
##                                                      Estimate Std. Error t value
## (Intercept)                                              53.6       41.8    1.28
## Age.Group0 to 4                                        -155.9       51.2   -3.04
## Age.Group5 to 9                                        -156.5       51.2   -3.06
## Age.Group10 to 14                                      -140.6       51.2   -2.75
## Age.Group15 to 19                                       259.8       51.2    5.07
## Age.Group20 to 24                                       390.9       51.2    7.63
## Age.Group25 to 29                                       135.2       51.2    2.64
## Age.Group35 to 39                                       -71.3       51.2   -1.39
## Age.Group40 to 44                                      -106.4       51.2   -2.08
## Age.Group45 to 54                                      -128.5       51.2   -2.51
## Age.Group55 to 64                                      -147.2       51.2   -2.87
## Age.Group65 or more                                    -154.7       51.2   -3.02
## EthnicityAmerican Indians & Alaskan Natives             68.3       33.1    2.06
## EthnicityAsians & Pacific Islanders                    -14.7       33.1   -0.44
## EthnicityHispanics                                      19.3       33.1    0.58
## EthnicityNon-Hispanic Blacks                           444.5       33.1   13.44
##                                                      Pr(>|t|)
## (Intercept)                                            0.2008
## Age.Group0 to 4                                        0.0024 **
## Age.Group5 to 9                                        0.0024 **
## Age.Group10 to 14                                      0.0062 **
## Age.Group15 to 19                                     5.3e-07 ***
## Age.Group20 to 24                                     9.4e-14 ***
## Age.Group25 to 29                                      0.0085 **
## Age.Group35 to 39                                      0.1647
## Age.Group40 to 44                                      0.0383 *
## Age.Group45 to 54                                      0.0124 *
## Age.Group55 to 64                                      0.0042 **
## Age.Group65 or more                                    0.0026 **
## EthnicityAmerican Indians & Alaskan Natives            0.0394 *
## EthnicityAsians & Pacific Islanders                    0.6576
## EthnicityHispanics                                     0.5603
## EthnicityNon-Hispanic Blacks                          < 2e-16 ***
## ---
## Signif. codes:  0 '***' 0.001 '**' 0.01 '*' 0.05 '.' 0.1 ' ' 1
##
## Residual standard error: 256 on 584 degrees of freedom
## Multiple R-squared:  0.491,  Adjusted R-squared:  0.477
## F-statistic: 37.5 on 15 and 584 DF,  p-value: <2e-16
```

The coefficients and p-values have changed because the reference point is now different, but there are still a lot of stars on the righthand side of the summary output, so we know that age and ethnicity have an impact on infection rates.

Plotting and Inspecting Models

lm models have a plot method that lets you check the goodness of fit in six different ways. In its simplest form, you can just call plot(*the_model*), and it draws several plots one after another. A slightly better approach is to use the layout function to see all the plots together, as demonstrated in Figure 15-2:

```
plot_numbers <- 1:6
layout(matrix(plot_numbers, ncol = 2, byrow = TRUE))
plot(model4, plot_numbers)
```

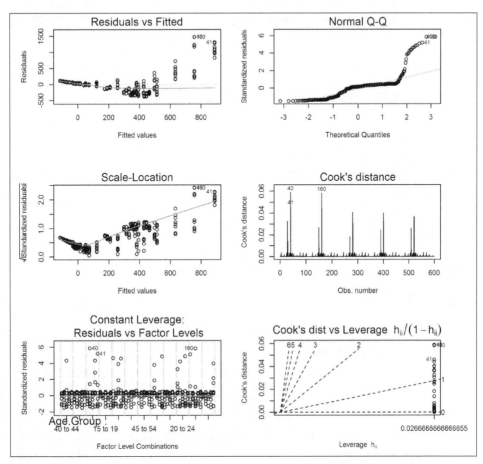

Figure 15-2. Diagnostic plots for a linear model

At the top end, there are some values that are too high (notice the large positive residuals at the righthand side of the "Residuals vs Fitted" plot, the points far above the line in the "Normal Q-Q" plot, and the spikes in the "Cook's distance" plot). In particular, rows 40, 41, and 160 have been singled out as outliers:

```
gonorrhoea[c(40, 41, 160), ]
```

```
##      Year Age.Group               Ethnicity Gender Rate
## 40   2007  15 to 19  Non-Hispanic Blacks Female 2239
## 41   2007  20 to 24  Non-Hispanic Blacks Female 2200
## 160  2008  15 to 19  Non-Hispanic Blacks Female 2233
```

These large values all refer to non-Hispanic black females, suggesting that we are perhaps missing an interaction term with ethnicity and gender.

The model variable that is returned by lm is a fairly complex beast. The output isn't included here for brevity, but you can explore the structure of these model variables in the usual way:

```
str(model4)
unclass(model4)
```

There are many convenience functions for accessing the various components of the model, such as formula, nobs, residuals, fitted, and coefficients:

```
formula(model4)

## Rate ~ Age.Group + Ethnicity
## <environment: 0x000000004ed4e110>

nobs(model4)

## [1] 600

head(residuals(model4))

##        1      2      3       4       5       6
##   102.61 102.93  87.25 -282.38 -367.61 -125.38

head(fitted(model4))

##        1      2      3      4      5      6
## -102.31 -102.93 -87.05 313.38 444.51 188.78

head(coefficients(model4))

##          (Intercept)   Age.Group0 to 4   Age.Group5 to 9 Age.Group10 to 14
##                53.56           -155.87           -156.49           -140.60
## Age.Group15 to 19 Age.Group20 to 24
##            259.83            390.95
```

Beyond these, there are more functions for diagnosing the quality of linear regression models (listed on the ?influence.measures page), and you can access the R^2 value ("the fraction of variance explained by the model") from the summary:

```
head(cooks.distance(model4))

##        1         2         3         4         5         6
## 0.0002824 0.0002842 0.0002042 0.0021390 0.0036250 0.0004217

summary(model4)$r.squared

## [1] 0.4906
```

These utility functions are great for providing alternative diagnostics. For example, if you don't want a base graphics plot of a model, you could roll your own ggplot2 version. Figure 15-3 shows an example plot of residuals versus fitted values:

```
diagnostics <- data.frame(
  residuals = residuals(model4),
  fitted    = fitted(model4)
)
ggplot(diagnostics, aes(fitted, residuals)) +
  geom_point() +
  geom_smooth(method = "loess")
```

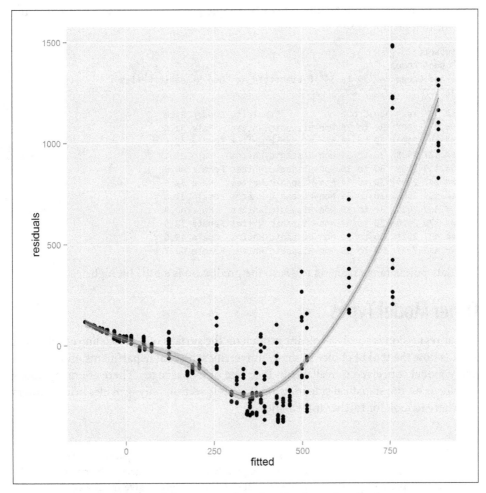

Figure 15-3. A ggplot2-based diagnostic plot of residuals vs. fitted values

The real beauty of using models is that you can predict outcomes based on whatever model inputs interest you. It's more fun when you have a time variable in the model so that you can predict the future, but in this case we can take a look at infection rates for specific demographics. In a completely self-interested fashion, I'm going to look at the infection rates for 30- to 34-year-old non-Hispanic white people:

```
new_data <- data.frame(
  Age.Group = "30 to 34",
  Ethnicity = "Non-Hispanic Whites"
)
predict(model4, new_data)

##     1
## 53.56
```

The model predicts an infection rate of 54 people per 100,000. Let's compare it to the data for that group:

```
subset(
  gonorrhoea,
  Age.Group == "30 to 34" & Ethnicity == "Non-Hispanic Whites"
)

##     Year Age.Group            Ethnicity Gender Rate
## 7   2007 30 to 34 Non-Hispanic Whites    Male 41.0
## 19  2007 30 to 34 Non-Hispanic Whites  Female 45.6
## 127 2008 30 to 34 Non-Hispanic Whites    Male 35.1
## 139 2008 30 to 34 Non-Hispanic Whites  Female 40.8
## 247 2009 30 to 34 Non-Hispanic Whites    Male 34.6
## 259 2009 30 to 34 Non-Hispanic Whites  Female 33.8
## 367 2010 30 to 34 Non-Hispanic Whites    Male 40.8
## 379 2010 30 to 34 Non-Hispanic Whites  Female 38.5
## 487 2011 30 to 34 Non-Hispanic Whites    Male 48.0
## 499 2011 30 to 34 Non-Hispanic Whites  Female 40.7
```

The data points range from 34 to 48, so the prediction is a little bit high.

Other Model Types

Linear regression is barely a hairline scratch on the surface of R's modeling capabilities. As R is now the tool of choice in many university statistical departments, more or less every model conceived is available in R or one of its packages. There are many books that focus on the statistical side of using R, so this section only provides brief pointers on where to look for further information.

 The modeling capabilities of R are spread across many packages written by many people, and consequently their syntax varies. The caret package provides wrappers to about 150 models to give them a consistent interface, along with some tools for model training and validation. Max Kuhn's *Applied Predictive Modeling* is the definitive reference to this package.

The lm function for linear (ordinary least squares regression) models has a generalization, glm, that lets you specify different distributions for error terms and transformations on the response variable. You can use it for logistic regression (where the response variable is logical or categorical), amongst other things, and the syntax is almost identical to that of `lm`:

```
glm(true_or_false ~ some + predictor + variables, data, family = binomial())
```

John Fox's *An R Companion to Applied Regression* is basically 472 pages of cool things you can do with the `glm` function.

The `nlme` package (which ships with R) contains the `lme` function for linear mixed-effects models and `nlme` for nonlinear mixed-effects models.[3] Again, the syntax is more or less the same as with `lm`:

```
lme(y ~ some + fixed + effects, data, random = ~ 1 | random / effects)
```

Mixed-Effects Models in S and S-PLUS by José Pinheiro and Doug Bates is the canonical reference, with loads of examples.

For response variables that are proportions, the `betareg` package contains a function of the same name that allows beta regression. This is especially important for Exercise 15-3.

For data miners (and others with high-dimensional datasets), the `rpart` package, which also ships with R, creates regression trees (a.k.a. decision trees). Even more excitingly, the `randomForest` package lets you create a whole forest of regression trees. `C50` and `mboost` provide gradient boosting, which many regard as even better than random forests.

`kmeans` lets you do k-means clustering, and there are several packages that provide specialist extensions, such as `kernlab` for weighted k-means, `kml` for longitudinal k-means, `trimclust` for trimmed k-means, and `skmeans` for spherical k-means.

If you do a lot of data mining, you may be interested in Rattle, a GUI that gives easy access to R's data-mining models. Graham Williams has written a book about it, *Data Mining with Rattle and R*, but start by visiting the website (*http://rattle.toga ware.com/*) to see if it takes your fancy.

Social scientists may appreciate traditional dimension-reduction models: factor analysis is supported through the `factanal` method, and principal components analysis has two methods (`princomp` provides S-Plus compatibility, and `prcomp` uses a more modern, numerically stable algorithm).

---

3. Mixed-effects models are regressions where some of your predictor variables affect the variance of the response rather than the mean. For example, if you measure oxygen levels in people's bloodstreams, you might want to know how much variation there is between people and between measurements for a person, but you don't care that oxygen levels are higher or lower for a specific person.

The deSolve package contains many methods for solving systems of ordinary/partial/ delay differential equations.

Summary

- You can generate random numbers from pretty much any distribution ever conceived.
- Most distributions also have functions available for calculating their PDF/CDF/ inverse CDF.
- Many modeling functions use formulae to specify the form of the model.
- lm runs a linear regression, and there are many support functions for updating, diagnosing, and predicting with these models.
- R can run a wide range of statistical models.

Test Your Knowledge: Quiz

Question 15-1
How would you create the same set of random numbers twice?

Question 15-2
What is the naming convention for PDF, CDF, and inverse CDF functions?

Question 15-3
What does a colon (:) mean in the context of model formulae?

Question 15-4
Which functions can you use to compare linear regression models?

Question 15-5
How do you determine the fraction of variance explained by a linear model?

Test Your Knowledge: Exercises

Exercise 15-1
1. While typing this book, I make about three typos per paragraph. Use the PDF of the Poisson distribution, dpois, to find the probability that I make exactly three typos in a given paragraph. [5]

2. A healthy 25-year-old woman having unprotected sex at the right time has a 25% chance of getting pregnant each month. Use the CDF for the negative binomial distribution, pnbinom, to calculate the probability that she will have become pregnant after a year. [5]

3. You need 23 people to have a 50% chance that two or more of them share the same birthday. Use the inverse CDF of the birthday distribution, qbirthday, to calculate how many people you need to have a 90% chance of a shared birthday. [5]

Exercise 15-2

Re-run the linear regression analysis on the gonorrhoea dataset, considering only 15- to 34-year-olds. Are the significant predictor variables different this time? [15]

For bonus points, explore adding interaction terms into the model. [15]

Exercise 15-3

Install and load the betareg package. Explore the obama_vs_mccain dataset using beta regression, via the betareg function in that package. Use the Obama column as the response variable.

To keep things simple, remove the "District of Columbia" outlier, don't bother with interactions, and only include one ethnicity and one religion column. (The ethnicity and religion columns aren't independent because they represent fractions of a total.) In the absence of political understanding, you may trust the p-values for the purposes of updating models. Hint: You need to rescale the Obama column to range from 0 to 1. [30]

## CHAPTER 16

Programming

Writing data analysis code is hard. This chapter is about what happens when things go wrong, and how to avoid that happening in the first place. We start with the different types of feedback that you can give to indicate a problem, working up to errors. Then we look at how to handle those errors when they are thrown, and how to debug code to eliminate the bad errors. A look at unit testing frameworks gives you the skills to avoid writing buggy code.

Next, we see some magic tricks: converting strings into code and code into strings ("Just like that!" as Tommy Cooper used to say). The chapter concludes with an introduction to some of the object-oriented programming systems in R.

Chapter Goals

After reading this chapter, you should:

- Know how to provide feedback to the user through messages, warnings, and errors
- Be able to gracefully handle errors
- Understand a few techniques for debugging code
- Be able to use the RUnit and testthat unit testing frameworks
- Know how to convert strings to R expressions and back again
- Understand the basics of the S3 and reference class object-oriented programming systems

Messages, Warnings, and Errors

We've seen the `print` function on many occasions for displaying variables to the console. For displaying diagnostic information about the state of the program, R has three functions. In increasing order of severity, they are `message`, `warning`, and `stop`.

`message` concatenates its inputs without spaces and writes them to the console. Some common uses are providing status updates for long-running functions, notifying users of new behavior when you've changed a function, and providing information on default arguments:

```
f <- function(x)
{
  message("'x' contains ", toString(x))
  x
}
f(letters[1:5])

## 'x' contains a, b, c, d, e

## [1] "a" "b" "c" "d" "e"
```

The main advantage of using `message` over `print` (or the lower-level `cat`) is that the user can turn off their display. It may seem trivial, but when you are repeatedly running the same code, not seeing the same message 100 times can have a wonderful effect on morale:

```
suppressMessages(f(letters[1:5]))

## [1] "a" "b" "c" "d" "e"
```

Warnings behave very similarly to messages, but have a few extra features to reflect their status as indicators of bad news. Warnings should be used when something has gone wrong, but not so wrong that your code should just give up. Common use cases are bad user inputs, poor numerical accuracy, or unexpected side effects:

```
g <- function(x)
{
  if(any(x < 0))
  {
    warning("'x' contains negative values: ", toString(x[x < 0]))
  }
  x
}
g(c(3, -7, 2, -9))

## Warning: 'x' contains negative values: -7, -9

## [1]  3 -7  2 -9
```

As with messages, warnings can be suppressed:

```
suppressWarnings(g(c(3, -7, 2, -9)))
## [1]  3 -7  2 -9
```

There is a global option, warn, that determines how warnings are handled. By default warn takes the value 0, which means that warnings are displayed when your code has finished running.

You can see the current level of the warn option using getOption:

```
getOption("warn")
## [1] 1
```

If you change this value to be less than zero, all warnings are ignored:

```
old_ops <- options(warn = -1)
g(c(3, -7, 2, -9))
## [1]  3 -7  2 -9
```

It is usually dangerous to completely turn off warnings, though, so you should reset the options to their previous state using:

```
options(old_ops)
```

Setting warn to 1 means that warnings are displayed as they occur, and a warn value of 2 or more means that all warnings are turned into errors.

You can access the last warning by typing last.warning.

I mentioned earlier that if the warn option is set to 0, then warnings are shown when your code finishes running. Actually, it's a little more complicated than that. If 10 or fewer warnings were generated, then this is what happens. But if there were more than 10 warnings, you get a message stating how many warnings were generated, and you have to type warnings() to see them. This is demonstrated in Figure 16-1.

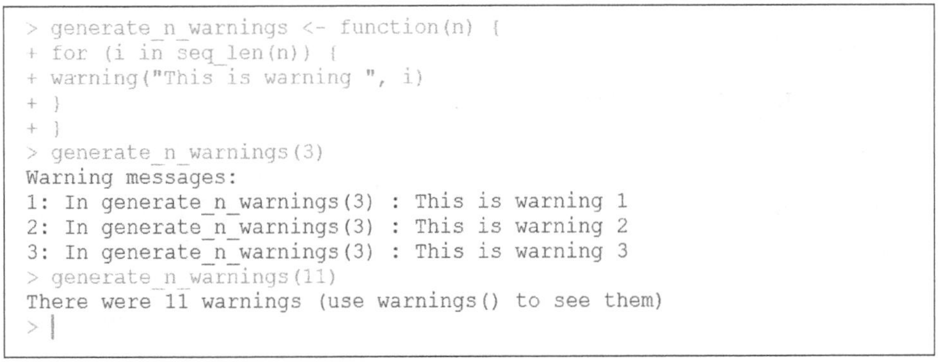

```
> generate_n_warnings <- function(n) {
+ for (i in seq_len(n)) {
+ warning("This is warning ", i)
+ }
+ }
> generate_n_warnings(3)
Warning messages:
1: In generate_n_warnings(3) : This is warning 1
2: In generate_n_warnings(3) : This is warning 2
3: In generate_n_warnings(3) : This is warning 3
> generate_n_warnings(11)
There were 11 warnings (use warnings() to see them)
> |
```

Figure 16-1. Where there are more than 10 warnings, use warnings to see them

Errors are the most serious condition, and throwing them halts further execution. Errors should be used when a mistake has occurred or you know a mistake will occur. Common reasons include bad user input that can't be corrected (by using an `as.*` function, for example), the inability to read from or write to a file, or a severe numerical error:

```
h <- function(x, na.rm = FALSE)
{
  if(!na.rm && any(is.na(x)))
  {
    stop("'x' has missing values.")
  }
  x
}
h(c(1, NA))

## Error: 'x' has missing values.
```

`stopifnot` throws an error if any of the expressions passed to it evaluate to something that isn't true. It provides a simple way of checking that the state of your program is as expected:

```
h <- function(x, na.rm = FALSE)
{
  if(!na.rm)
  {
    stopifnot(!any(is.na(x)))
  }
  x
}
h(c(1, NA))

## Error: !any(is.na(x)) is not TRUE
```

For a more extensive set of human-friendly tests, use the `assertive` package:

```
library(assertive)
h <- function(x, na.rm = FALSE)
{
  if(!na.rm)
  {
    assert_all_are_not_na(x)
  }
  x
}
h(c(1, NA))

## Error: x contains NAs.
```

Error Handling

Some tasks are inherently risky. Reading from and writing to files or databases is notoriously error prone, since you don't have complete control over the filesystem, or the network or database. In fact, any time that R interacts with other software (Java code via rJava, WinBUGS via R2WinBUGS, or any of the hundreds of other pieces of software that R can connect to), there is an inherent risk that something will go wrong.

For these dangerous tasks,[1] you need to decide what to do when problems occur. Sometimes it isn't useful to stop execution when an error is thrown. For example, if you are looping over files importing them, then if one import fails you don't want to just stop executing and lose all the data that you've successfully imported already.

In fact, this point generalizes: any time you are doing something risky in a loop, you don't want to discard your progress if one of the iterations fails. In this next example, we try to convert each element of a list into a data frame:

```
to_convert <- list(
  first  = sapply(letters[1:5], charToRaw),
  second = polyroot(c(1, 0, 0, 0, 1)),
  third  = list(x = 1:2, y = 3:5)
)
```

If we run the code nakedly, it fails:

```
lapply(to_convert, as.data.frame)

## Error: arguments imply differing number of rows: 2, 3
```

Oops! The third element fails to convert because of differing element lengths, and we lose everything.

The simplest way of protecting against total failure is to wrap the failure-prone code inside a call to the try function:

```
result <- try(lapply(to_convert, as.data.frame))
```

Now, although the error will be printed to the console (you can suppress this by passing silent = TRUE), execution of code won't stop.

If the code passed to a try function executes successfully (without throwing an error), then result will just be the result of the calculation, as usual. If the code fails, then result will be an object of class try-error. This means that after you've written a line of code that includes try, the next line should always look something like this:

---

1. OK, connecting to a file isn't wrestling an angry bear, but it's high-risk in programming terms.

```
if(inherits(result, "try-error"))
{
  #special error handling code
} else
{
  #code for normal execution
}
## NULL
```

Since you have to include this extra line every time, code using the try function is a bit ugly. A prettier alternative[2] is to use tryCatch. tryCatch takes an expression to safely run, just as try does, but also has error handling built into it.

To handle an error, you pass a function to an argument named error. This error argument accepts an error (technically, an object of class simpleError) and lets you manipulate, print, or ignore it as you see fit. If this sounds complicated, don't worry: it's easier in practice. In this next example, when an error is thrown, we print the error message and return an empty data frame:

```
tryCatch(
  lapply(to_convert, as.data.frame),
  error = function(e)
  {
    message("An error was thrown: ", e$message)
    data.frame()
  }
)
## An error was thrown: arguments imply differing number of rows: 2, 3

## data frame with 0 columns and 0 rows
```

tryCatch has one more trick: you can pass an expression to an argument named finally, which runs whether an error was thrown or not (just like the on.exit function we saw when we were connecting to databases).

Despite having played with try and tryCatch, we still haven't solved our problem: when looping over things, if an error is thrown, we want to keep the results of the iterations that worked.

To achieve this, we need to put try or tryCatch inside the loop:

2. Don't underestimate the importance of pretty code. You'll spend more time reading code than writing it.

```
lapply(
  to_convert,
  function(x)
  {
    tryCatch(
      as.data.frame(x),
      error = function(e) NULL
    )
  }
)
## $first
##    x
## a 61
## b 62
## c 63
## d 64
## e 65
##
## $second
##                 x
## 1  0.7071+0.7071i
## 2 -0.7071+0.7071i
## 3 -0.7071-0.7071i
## 4  0.7071-0.7071i
##
## $third
## NULL
```

Since this is a common piece of code, the plyr package contains a function, tryapply, that deals with exactly this case in a cleaner fashion:

```
tryapply(to_convert, as.data.frame)
## $first
##    x
## a 61
## b 62
## c 63
## d 64
## e 65
##
## $second
##                 x
## 1  0.7071+0.7071i
## 2 -0.7071+0.7071i
## 3 -0.7071-0.7071i
## 4  0.7071-0.7071i
```

Eagle-eyed observers may notice that the failures are simply removed in this case.

Debugging

All nontrivial software contains errors.[3] When problems happen, you need to be able to find where they occur, and hopefully find a way to fix them. This is especially true if it's your own code. If the problem occurs in the middle of a simple script, you usually have access to all the variables, so it is trivial to locate the problem.

More often than not, problems occur somewhere deep inside a function inside another function inside another function. In this case, you need a strategy to inspect the state of the program at each level of the call stack. ("Call stack" is just jargon for the list of functions that have been called to get you to this point in the code.)

When an error is thrown, the `traceback` function tells you where the last error occurred. First, let's define some functions in which the error can occur:

```
outer_fn <- function(x) inner_fn(x)
inner_fn <- function(x) exp(x)
```

Now let's call `outer_fn` (which then calls `inner_fn`) with a bad input:

```
outer_fn(list(1))
```

```
## Error: non-numeric argument to mathematical function
```

`traceback` now tells us the functions that we called before tragedy struck (see Figure 16-2).

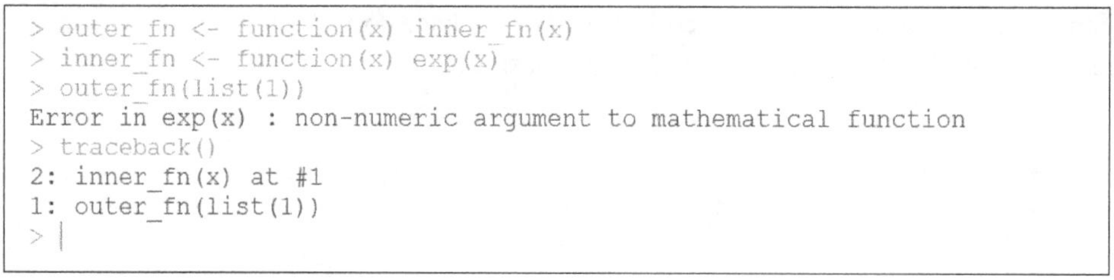

```
> outer_fn <- function(x) inner_fn(x)
> inner_fn <- function(x) exp(x)
> outer_fn(list(1))
Error in exp(x) : non-numeric argument to mathematical function
> traceback()
2: inner_fn(x) at #1
1: outer_fn(list(1))
> |
```

Figure 16-2. Call stack using traceback

In general, if it isn't an obvious bug, we don't know where in the call stack the problem occurred. One reasonable strategy is to start in the function where the error was thrown, and work our way up the stack if we need to. To do this, we need a way to stop execution of the code close to the point where the error was thrown. One way to do this is to add a call to the `browser` function just before the error point (we know where the error occurred because we used `traceback`):

---

3. Space shuttle software was reputed to contain just one bug in 420,000 lines of code (*http://www.fastcompany.com/28121/they-write-right-stuff*), but that level of formal development methodology, code peer-reviewing, and extensive testing doesn't come cheap.

```
inner_fn <- function(x)
{
  browser()      #execution pauses here
  exp(x)
}
```

browser halts execution when it is reached, giving us time to inspect the program. A really good idea in most cases is to call ls.str to see the values of all the variables that are in play at the time. In this case we see that x is a list, not a numeric vector, causing exp to fail.

An alternative strategy for spotting errors is to set the global error option. This strategy is preferable when the error lies inside someone else's package, where it is harder to stick a call to browser. (You can alter functions inside installed packages using the fixInNamespace function. The changes persist until you close R.)

The error option accepts a function with no arguments, and is called whenever an error is thrown. As a simple example, we can set it to print a message after the error has occurred, as shown in Figure 16-3.

```
> oh_dear <- function() message("Oh dear!")
> old_ops <- options(error = oh_dear)
> stop("I will break your program!")
Error: I will break your program!
Oh dear!
> |
```

Figure 16-3. Overriding the global error option

While a sympathetic message may provide a sliver of consolation for the error, it isn't very helpful in terms of fixing the problem. A much more useful alternative is provided in the recover function that ships with R. recover lets you step into any function in the call stack after an error has been thrown (see Figure 16-4).

```
> outer_fn(list(1))
Error in exp(x) : non-numeric argument to mathematical function

Enter a frame number, or 0 to exit

1: outer_fn(list(1))
2: #1: inner_fn(x)

Selection: |
```

Figure 16-4. Call stack using error = recover

You can also step through a function line by line using the debug function. This is a bit boring with trivial single-line functions like inner and outer, so we'll test it on a more substantial offering. buggy_count, included in the learningr package, is a buggy version of the count function from the plyr package that fails in an obscure way when you pass it a factor. Pressing Enter at the command line without typing anything lets us step through it until we find the problem:

```
debug(buggy_count)
x <- factor(sample(c("male", "female"), 20, replace = TRUE))
buggy_count(x)
```

count (and by extension, our buggy_count) accepts a data frame or a vector as its first argument. If the df argument is a vector, then the function inserts it into a data frame.

Figure 16-5 shows what happens when we reach this part of the code. When df is a factor, we want it to be placed inside a data frame. Unfortunately, is.vector returns FALSE for factors, and the step is ignored. Factors aren't considered to be vectors, because they have attributes other than names. What the code really should contain (and does in the proper version of plyr) is a call to is.atomic, which is TRUE for factors as well as other vector types, like numeric.

```
Browse[2]>
debug at #3: if (is.vector(df)) {
    df <- data.frame(x = df)
}
Browse[2]> is.vector(df)
[1] FALSE
Browse[2]> is.atomic(df)
[1] TRUE
```

Figure 16-5. Debugging the buggy_count function

To exit the debugger, type Q at the command line. With the debug function, the debugger will be started every time that function is called. To turn off debugging, call undebug:

```
undebug(buggy_count)
```

As an alternative, use debugonce, which only calls the debugger the first time a function is called.[4]

Testing

To make sure that your code isn't buggy and awful, it is important to test it. *Unit testing* is the concept of testing small chunks of code; in R this means testing at the functional

---

4. As Tobias Verbeke of Open Analytics once quipped, "debugonce is a very optimistic function. I think de bugtwice might have been better."

level. (*System* or *integration testing* is the larger-scale testing of whole pieces of software, but that is more useful for application development than data analysis.)

Each time you change a function, you can break code in other functions that rely on it. This means that each time you change a function, you need to test everything that it could affect. Attempted manually, this is impossible, or at least time-consuming and boring enough that you won't do it. Consequently, you need to automate the task. In R, you have two choices for this:

1. RUnit has "xUnit" syntax, meaning that it's very similar to Java's JUnit, .NET's NUnit, Python's PyUnit, and a whole other family of unit testing suites. This makes it easiest to learn if you've done unit testing in any other language.

2. testthat has its own syntax, and a few extra features. In particular, the caching of tests makes it much faster for large projects.

Let's test the hypotenuse function we wrote when we first learned about functions in "Functions" on page 82. It uses the obvious algorithm that you might use for pen and paper calculations.[5] The function is included in the learningr package:

```
hypotenuse <- function(x, y)
{
  sqrt(x ^ 2 + y ^ 2)
}
```

RUnit

In RUnit, each test is a function that takes no inputs. Each test compares the actual result of running some code (in this case, calling hypotenuse) to an expected value, using one of the check* functions contained in the package. In this next example we use checkEqualsNumeric, since we are comparing two numbers:

```
library(RUnit)
test.hypotenuse.3_4.returns_5 <- function()
{
  expected <- 5
  actual <- hypotenuse(3, 4)
  checkEqualsNumeric(expected, actual)
}
```

---

5. If you've just recoiled in horror at the phrase "pen and paper calculations," congratulations! You are well on your way to becoming an R user.

There is no universal naming convention for tests, but RUnit looks for functions with names beginning with test by default. The convention used here is designed to maximize clarity. Tests take the name form of test.*name_of_function.description_of_inputs.re turns_a_value.*

Sometimes we want to make sure that a function fails in the correct way. For example, we can test that hypotenuse fails if no inputs are provided:

```
test.hypotenuse.no_inputs.fails <- function()
{
  checkException(hypotenuse())
}
```

Many algorithms suffer loss of precision when given very small or very large inputs, so it is good practice to test those conditions. The smallest and largest positive numeric values that R can represent are given by the double.xmin and double.xmax components of the built-in .Machine constant:

```
.Machine$double.xmin
```

```
## [1] 2.225e-308
```

```
.Machine$double.xmax
```

```
## [1] 1.798e+308
```

For the small and large tests, we pick values close to these limits. In the case of small numbers, we need to manually tighten the tolerance of the test. By default, checkE qualsNumeric considers its test passed when the actual result is within about 1e-8 of the expected result (it uses absolute, not relative differences). We set this value to be a few orders of magnitude smaller than the inputs to make sure that the test fails appropriately:

```
test.hypotenuse.very_small_inputs.returns_small_positive <- function()
{
  expected <- sqrt(2) * 1e-300
  actual <- hypotenuse(1e-300, 1e-300)
  checkEqualsNumeric(expected, actual, tolerance = 1e-305)
}
test.hypotenuse.very_large_inputs.returns_large_finite <- function()
{
  expected <- sqrt(2) * 1e300
  actual <- hypotenuse(1e300, 1e300)
  checkEqualsNumeric(expected, actual)
}
```

There are countless more possible tests; for example, what happens if we pass missing values or NULL or infinite values or character values or vectors or matrices or data frames, or we expect an answer in non-Euclidean space? Thorough testing works your imagination hard. Unleash your inner two-year-old and contemplate breaking stuff. On this occasion, we'll stop here. Save all your tests into a file; RUnit defaults to looking for files with names that begin with "runit" and have a *.R* file extension. These tests can be found in the *tests* directory of the learningr package.

Now that we have some tests, we need to run them. This is a two-step process.

First, we define a test suite with defineTestSuite. This function takes a string for a name (used in its output), and a path to the directory where your tests are contained. If you've named your test functions or files in a nonstandard way, you can provide a pattern to identify them:

```
test_dir <- system.file("tests", package = "learningr")
suite <- defineTestSuite("hypotenuse suite", test_dir)
```

The second step is to run them with runTestSuite (additional line breaks have been added here as needed, to fit the formatting of the book):

```
runTestSuite(suite)

##
##
## Executing test function test.hypotenuse.3_4.returns_5  ...
## done successfully.
##
##
##
## Executing test function test.hypotenuse.no_inputs.fails  ...
## done successfully.
##
##
##
## Executing test function
## test.hypotenuse.very_large_inputs.returns_large_finite  ...
## Timing stopped at: 0 0 0 done successfully.
##
##
##
## Executing test function
## test.hypotenuse.very_small_inputs.returns_small_positive  ...
## Timing stopped at: 0 0 0 done successfully.

## Number of test functions: 4
## Number of errors: 0
## Number of failures: 2
```

This runs each test that it finds and displays whether it passed, failed, or threw an error. In this case, you can see that the small and large input tests failed. So what went wrong?

The problem with our algorithm is that we have to square each input. Squaring big numbers makes them larger than the largest (double-precision) number that R can represent, so the result comes back as infinity. Squaring very small numbers makes them even smaller, so that R thinks they are zero. (There are better algorithms that avoid this problem; see the ?hypotenuse help page for links to a discussion of better algorithms for real-world use.)

RUnit has no built-in checkWarning function to test for warnings. To test that a warning has been thrown, we need a trick: we set the warn option to 2 so that warnings become errors, and then restore it to its original value when the test function exits using on.exit. Recall that code inside on.exit is run when a function exits, regardless of whether it completed successfully or an error was thrown:

```
test.log.minus1.throws_warning <- function()
{
  old_ops <- options(warn = 2) #warnings become errors
  on.exit(old_ops)             #restore old behavior
  checkException(log(-1))
}
```

testthat

Though testthat has a different syntax, the principles are almost the same. The main difference is that rather than each test being a function, it is a call to one of the expect_* functions in the package. For example, expect_equal is the equivalent of RUnit's check EqualsNumeric. The translated tests (also available in the *tests* directory of the lear ningr package) look like this:

```
library(testthat)
expect_equal(hypotenuse(3, 4), 5)
expect_error(hypotenuse())
expect_equal(hypotenuse(1e-300, 1e-300), sqrt(2) * 1e-300, tol = 1e-305)
expect_equal(hypotenuse(1e300, 1e300), sqrt(2) * 1e300)
```

To run this, we call test_file with the name of the file containing tests, or test_dir with the name of the directory containing the files containing the tests. Since we have only one file, we'll use test_file:

```
filename <- system.file(
  "tests",
  "testthat_hypotenuse_tests.R",
  package = "learningr"
)
test_file(filename)

## ..12
##
## 1. Failure: (unknown) --------------------------------------------------
## learningr::hypotenuse(1e-300, 1e-300) not equal to sqrt(2) * 1e-300
```

```
## Mean relative difference: 1
##
## 2. Failure: (unknown) ---------------------------------------------------
## learningr::hypotenuse(1e+300, 1e+300) not equal to sqrt(2) * 1e+300
## Mean relative difference: Inf
```

There are two variations for running the tests: `test_that` tests code that you type at the command line (or, more likely, copy and paste), and `test_package` runs all tests from a package, making it easier to test nonexported functions.

Unlike with RUnit, warnings can be tested for directly via `expect_warning`:

```
expect_warning(log(-1))
```

Magic

The source code that we write, as it exists in a text editor, is just a bunch of strings. When we run that code, R needs to interpret what those strings contain and perform the appropriate action. It does that by first turning the strings into one of several *language* variable types. And sometimes we want to do the opposite thing, converting language variables into strings.

Both these tasks are rather advanced, dark magic. As is the case with magic in every movie ever, if you use it without understanding what you are doing, you'll inevitably suffer nasty, unexpected consequences. On the other hand, used knowledgeably and sparingly, there are some useful tricks that you can put up your sleeve.

Turning Strings into Code

Whenever you type a line of code at the command line, R has to turn that string into something it understands. Here's a simple call to the arctangent function:

```
atan(c(-Inf, -1, 0, 1, Inf))
## [1] -1.5708 -0.7854  0.0000  0.7854  1.5708
```

We can see what happens to this line of code in slow motion by using the `quote` function. `quote` takes a function call like the one in the preceding line, and returns an object of class `call`, which represents an "unevaluated function call":

```
(quoted_r_code <- quote(atan(c(-Inf, -1, 0, 1, Inf))))
## atan(c(-Inf, -1, 0, 1, Inf))
class(quoted_r_code)
## [1] "call"
```

The next step that R takes is to evaluate that call. We can mimic this step using the `eval` function:

```
eval(quoted_r_code)

## [1] -1.5708 -0.7854  0.0000  0.7854  1.5708
```

The general case, then, is that to execute code that you type, R does something like
`eval(quote(`*the stuff you typed at the command line*`))`.

To understand the `call` type a little better, let's convert it to a list:

```
as.list(quoted_r_code)

## [[1]]
## atan
##
## [[2]]
## c(-Inf, -1, 0, 1, Inf)
```

The first element is the function that was called, and any additional elements contain
the arguments that were passed to it.

One important thing to remember is that in R, more or less everything is a function.
That's a slight exaggeration, but operators like +; language constructs like `switch`, `if`,
and `for`; and assignment and indexing are functions:

```
vapply(
  list(`+`, `if`, `for`, `<-`, `[`, `[[`),
  is.function,
  logical(1)
)

## [1] TRUE TRUE TRUE TRUE TRUE TRUE
```

The upshot of this is that anything that you type at the command line is really a function
call, which is why this input is turned into `call` objects.

All of this was a long-winded way of saying that sometimes we want to take text that is
R code and get R to execute it. In fact, we've already seen two functions that do exactly
that for special cases: `assign` takes a string and assigns a value to a variable with that
name, and its reverse, `get`, retrieves a variable based upon a string input.

Rather than just limiting ourselves to assigning and retrieving variables, we might occasionally decide that we want to take an arbitrary string of R code and execute it. You
may have noticed that when we use the `quote` function, we just type the R code directly
into it, without wrapping it in—ahem—quotes. If our input is a string (as in a character
vector of length one), then we have a slightly different problem: we must "parse" the
string. Naturally, this is done with the `parse` function.

`parse` returns an `expression` object rather than a call. Before you get frightened, note
that an `expression` is basically just a list of calls.

The exact nature of calls and expressions is deep, dark magic, and I don't want to be responsible for the ensuing zombie apocalypse when you try to raise the dead using R. If you are interested in arcana, read Chapter 6 of the R Language Definition manual that ships with R.

When we call parse in this way, we must explicitly name the text argument:

```
parsed_r_code <- parse(text = "atan(c(-Inf, -1, 0, 1, Inf))")
class(parsed_r_code)
```

```
## [1] "expression"
```

Just as with the quoted R code, we use eval to evaluate it:

```
eval(parsed_r_code)
```

```
## [1] -1.5708 -0.7854  0.0000  0.7854  1.5708
```

 This sort of mucking about with evaluating strings is a handy trick, but the resulting code is usually fragile and fiendish to debug, making your code unmaintainable. This is the zombie (code) apocalypse mentioned above.

Turning Code into Strings

There are a few occasions when we want to solve the opposite problem: turning code into a string. The most common reason for this is to use the name of a variable that was passed into a function. The base histogram-drawing function, hist, includes a default title that tells you the name of the data variable:

```
random_numbers <- rt(1000, 2)
hist(random_numbers)
```

To replicate this technique ourselves, we need two functions: substitute and deparse. substitute takes some code and returns a language object. That usually means a call, like we would have created using quote, but occasionally it's a name object, which is a special type that holds variable names. (Don't worry about the details, this section is called "Magic" for a reason.)

The next step is to turn this language object into a string. This is called *deparsing*. The technique can be very useful for providing helpful error messages when you check user inputs to functions. Let's see the deparse-substitute combination in action:

```
divider <- function(numerator, denominator)
{
  if(denominator == 0)
  {
    denominator_name <- deparse(substitute(denominator))
    warning("The denominator, ", sQuote(denominator_name), ", is zero.")
  }
  numerator / denominator
}
top <- 3
bottom <- 0
divider(top, bottom)
## Warning: The denominator, 'bottom', is zero.

## [1] Inf
```

substitute has one more trick up its sleeve when used in conjunction with eval. eval lets you pass it an environment or a data frame, so you can tell R where to look to evaluate the expression.

As a simple example, we can use this trick to retrieve the levels of the Gender column of the hafu dataset:

```
eval(substitute(levels(Gender)), hafu)
## [1] "F" "M"
```

This is exactly how the with function works:

```
with(hafu, levels(Gender))
## [1] "F" "M"
```

In fact, many functions use the technique: subset uses it in several places, and lattice plots use the trick to parse their formulae. There are a few variations on the trick described in Thomas Lumley's "Standard nonstandard evaluation rules."

Object-Oriented Programming

Most R code that we've seen so far is functional-programming-inspired imperative programming. That is, functions are first-class objects, but we usually end up with a data-analysis script that executes one line at a time.

In a few circumstances, it is useful to use an object-oriented programming (OOP) style. This means that data is stored inside a class along with the functions that are allowed to act on it. It is an excellent tool for managing complexity in larger programs, and is particularly suited to GUI development (in R or elsewhere). See Michael Lawrence's *Programming Graphical User Interfaces in R* for more on that topic.

R has six (count 'em) different OOP systems, but don't let that worry you—there are only two of them that you'll need for new projects.

Three systems are built into R:

1. *S3* is a lightweight system for overloading functions (i.e., calling a different version of the function depending upon the type of input).

2. *S4* is a fully featured OOP system, but it's clunky and tricky to debug. Only use it for legacy code.

3. *Reference classes* are the modern replacement for S4 classes.

Three other systems are available in add-on packages (but for new code, you will usually want to use reference classes instead):

1. `proto` is a lightweight wrapper around environments for prototype-based programming.

2. `R.oo` extends S3 into a fully fledged OOP system.

3. `OOP` is a precursor to reference classes, now defunct.

 In many object-oriented programming languages, functions are called *methods*. In R, the two words are interchangeable, but "method" is often used in an OOP context.

S3 Classes

Sometimes we want a function to behave differently depending upon the type of input. A classic example is the `print` function, which gives a different style of output for different variables. S3 lets us call a different function for printing different kinds of variables, without having to remember the names of each one.

The `print` function is very simple—just one line, in fact:

```
print
## function (x, ...)
## UseMethod("print")
## <bytecode: 0x0000000018fad228>
## <environment: namespace:base>
```

It takes an input, x (and `...`; the ellipsis is necessary), and calls `UseMethod("print")`. `UseMethod` checks the class of x and looks for another function named `print.class_of_x`, calling it if it is found. If it can't find a function of that name, it tries to call `print.default`.

For example, if we want to print a `Date` variable, then we can just type:

```
today <- Sys.Date()
print(today)

## [1] "2013-07-17"
```

print calls the Date-specific function print.Date:

```
print.Date

## function (x, max = NULL, ...)
## {
##     if (is.null(max))
##         max <- getOption("max.print", 9999L)
##     if (max < length(x)) {
##         print(format(x[seq_len(max)]), max = max, ...)
##         cat(" [ reached getOption(\"max.print\") -- omitted",
##             length(x) - max, "entries ]\n")
##     }
##     else print(format(x), max = max, ...)
##     invisible(x)
## }
## <bytecode: 0x0000000006dc19f0>
## <environment: namespace:base>
```

Inside print.Date, our date is converted to a character vector (via format), and then print is called again. There is no print.character function, so this time UseMethod delegates to print.default, at which point our date string appears in the console.

 If a class-specific method can't be found, and there is no default method, then an error is thrown.

You can see all the available methods for a function with the methods function. The print function has over 100 methods, so here we just show the first few:

```
head(methods(print))

## [1] "print.abbrev"      "print.acf"       "print.AES"
## [4] "print.anova"       "print.Anova"     "print.anova.loglm"

methods(mean)

## [1] mean.Date     mean.default  mean.difftime mean.POSIXct  mean.POSIXlt
## [6] mean.times*   mean.yearmon* mean.yearqtr* mean.zoo*
##
##     Non-visible functions are asterisked
```

If you use dots in your function names, like data.frame, then it can get confusing as to which S3 method gets called. For example, print.data.frame could mean a print.data method for a frame input, as well as the correct sense of a print method for a data.frame object. Consequently, using lower_under_case or lowerCamelCase is preferred for new function names.

Reference Classes

Reference classes are closer to a classical OOP system than S3 and S4, and should be moderately intuitive to anyone who has used classes in C++ or its derivatives.

A *class* is the general template for how the variables should be structured. An *object* is a particular instance of the class. For example, 1:10 is an *object* of *class* numeric.

The setRefClass function creates the template for a class. In R terminology, it's called a *class generator*. In some other languages, it would be called a *class factory*.

Let's try to build a class for a 2D point as an example. A call to setRefClass looks like this:

```
my_class_generator <- setRefClass(
  "MyClass",
  fields = list(
    #data variables are defined here
  ),
  methods = list(
    #functions to operate on that data go here
    initialize = function(...)
    {
    #initialize is a special function called
    #when an object is created.
    }
  )
)
```

Our class needs x and y coordinates to store its location, and we want these to be numeric.

In the following example, we declare x and y to be numeric:

If we didn't care about the class of x and y, we could declare them with the special value ANY.

```
point_generator <- setRefClass(
  "point",
  fields = list(
    x = "numeric",
    y = "numeric"
  ),
  methods = list(
    #TODO
  )
)
```

This means that if we try to assign them values of another type, an error will be thrown. Purposely restricting user input may sound counterintuitive, but it can save you from having more obscure bugs further down the line.

Next we need to add an `initialize` method. This is called every time we create a `point` object. This method takes x and y input numbers and assigns them to our x and y fields. There are three interesting things to note about it:

1. If the first line of a method is a string, then it is considered to be help text for that method.

2. The global assignment operator, `<<-`, is used to assign to a field. Local assignment (using `<-`) just creates a local variable inside the method.

3. It is best practice to let `initialize` work without being passed any arguments, since it makes inheritance easier, as we'll see in a moment. This is why the x and y arguments have default values.[6]

With the `initialize` method, our class generator now looks like this:

```
point_generator <- setRefClass(
  "point",
  fields = list(
    x = "numeric",
    y = "numeric"
  ),
  methods = list(
    initialize = function(x = NA_real_, y = NA_real_)
    {
      "Assign x and y upon object creation."
      x <<- x
      y <<- y
    }
  )
)
```

6. In case you were wondering, `NA_real_` is a missing number. Usually for missing values we just use `NA` and let R figure out the type that it needs to be, but in this case, because we specified that the fields must be numeric, we need to explicitly state the type.

Our point class generator is finished, so we can now create a `point` object. Every generator has a new method for this purpose. The new method calls `initialize` (if it exists) as part of the object creation process:

```
(a_point <- point_generator$new(5, 3))
## Reference class object of class "point"
## Field "x":
## [1] 5
## Field "y":
## [1] 3
```

Generators also have a `help` method that returns the help string for a method that you specify:

```
point_generator$help("initialize")
## Call:
## $initialize(x = , y = )
##
##
## Assign x and y upon object creation.
```

You can provide a more traditional interface to object-oriented code by wrapping class methods inside other functions. This can be useful if you want to distribute your code to other people without having to teach them about OOP:

```
create_point <- function(x, y)
{
  point_generator$new(x, y)
}
```

At the moment, the class isn't very interesting because it doesn't do anything. Let's redefine it with some more methods:

```
point_generator <- setRefClass(
  "point",
  fields  = list(
    x = "numeric",
    y = "numeric"
  ),
  methods = list(
    initialize        = function(x = NA_real_, y = NA_real_)
    {
      "Assign x and y upon object creation."
      x <<- x
      y <<- y
    },
    distanceFromOrigin = function()
    {
      "Euclidean distance from the origin"
      sqrt(x ^ 2 + y ^ 2)
    },
```

```
  add                 = function(point)
  {
    "Add another point to this point"
    x <<- x + point$x
    y <<- y + point$y
    .self
  }
  )
)
```

These additional methods belong to point objects, unlike new and help, which belong to the class generator (in OOP terminology, new and help are *static* methods):

```
a_point <- create_point(3, 4)
a_point$distanceFromOrigin()

## [1] 5

another_point <- create_point(4, 2)
(a_point$add(another_point))

## Reference class object of class "point"
## Field "x":
## [1] 7
## Field "y":
## [1] 6
```

As well as new and help, generator classes have a few more methods. fields and methods respectively list the fields and methods of that class, and lock makes a field read-only:

```
point_generator$fields()

##          x          y
## "numeric" "numeric"

point_generator$methods()

##  [1] "add"               "callSuper"        "copy"
##  [4] "distanceFromOrigin" "export"           "field"
##  [7] "getClass"          "getRefClass"      "import"
## [10] "initFields"        "initialize"       "show"
## [13] "trace"             "untrace"          "usingMethods"
```

Some other methods can be called either from the generator object or from instance objects. show prints the object, trace and untrace let you use the trace function on a method, export converts the object to another class type, and copy makes a copy.

Reference classes support inheritance, where classes can have children to extend their functionality. For example, we can create a three-dimensional point class that contains our original point class, but includes an extra z coordinate.

A class inherits fields and methods from another class by using the `contains` argument:

```
three_d_point_generator <- setRefClass(
  "three_d_point",
  fields   = list(
    z = "numeric"
  ),
  contains = "point",          #this line lets us inherit
  methods  = list(
    initialize = function(x, y, z)
    {
      "Assign x and y upon object creation."
      x <<- x
      y <<- y
      z <<- z
    }
  )
)
a_three_d_point <- three_d_point_generator$new(3, 4, 5)
```

At the moment, our `distanceFromOrigin` function is wrong, since it doesn't take the z dimension into account:

```
a_three_d_point$distanceFromOrigin() #wrong!

## [1] 5
```

We need to override it in order for it to make sense in the new class. This is done by adding a method with the same name to the class generator:

```
three_d_point_generator <- setRefClass(
  "three_d_point",
  fields   = list(
    z = "numeric"
  ),
  contains = "point",
  methods  = list(
    initialize = function(x, y, z)
    {
      "Assign x and y upon object creation."
      x <<- x
      y <<- y
      z <<- z
    },
    distanceFromOrigin = function()
    {
      "Euclidean distance from the origin"
      sqrt(x ^ 2 + y ^ 2 + z ^ 2)
    }
  )
)
```

To use the updated definition, we need to recreate our point:

```
a_three_d_point <- three_d_point_generator$new(3, 4, 5)
a_three_d_point$distanceFromOrigin()
```

```
## [1] 7.071
```

Sometimes we want to use methods from the parent class (a.k.a. *superclass*). The `call Super` method does exactly this, so we could have written our 3D `distanceFromOri gin` (inefficiently) like this:

```
distanceFromOrigin = function()
{
  "Euclidean distance from the origin"
  two_d_distance <- callSuper()
  sqrt(two_d_distance ^ 2 + z ^ 2)
}
```

OOP is a big topic, and even limited to reference classes, it's worth a book in itself. John Chambers (creator of the S language, R Core member, and author of the reference classes code) is currently writing a book on OOP in R. Until that materializes, the `?Reference Classes` help page is currently the definitive reference-class reference.

Summary

- R has three levels of feedback about problems: messages, warnings, and errors.
- Wrapping code in a call to `try` or `tryCatch` lets you control how you handle errors.
- The `debug` function and its relatives help you debug functions.
- The `RUnit` and `testthat` packages let you do unit testing.
- R code consists of language objects known as calls and expressions.
- You can turn strings into language objects and vice versa.
- There are six different object-oriented programming systems in R, though only S3 and reference classes are needed for new projects.

Test Your Knowledge: Quiz

Question 16-1
    If your code generates more than 10 warnings, which function would you use to view them?

Question 16-2
    What is the class of the return value from a call to `try` if an error was thrown?

310 | Chapter 16: Programming

Question 16-3

To test that an error is correctly thrown in an RUnit test, you call the `checkExcep` `tion` function. What is the `testthat` equivalent?

Question 16-4

Which two functions do you need to mimic executing code typed at the command line?

Question 16-5

How do you make the `print` function do different things for different types of input?

Test Your Knowledge: Exercises

Exercise 16-1

The harmonic mean is defined as the reciprocal of the arithmetic mean of the reciprocal of the data, or `1 / mean(1 / x)`, where x contains positive numbers. Write a harmonic mean function that gives appropriate feedback when the input is not numeric or contains nonpositive values. [10]

Exercise 16-2

Using either `RUnit` or `testthat`, write some tests for your harmonic mean function. You should check that the harmonic mean of 1, 2, and 4 equals `12 / 7`; that passing no inputs throws an error; that passing missing values behaves correctly; and that it behaves as you intended for nonnumeric and nonpositive inputs. Keep testing until all the tests pass! [15]

Exercise 16-3

Modify your harmonic mean function so that the return value has the class `harmonic`. Now write an S3 `print` method for this class that displays the message "The harmonic mean is y," where y is the harmonic mean. [10]

## CHAPTER 17

Making Packages

R's success lies in its community. While the R Core Team does a fantastic job, it's important to realize that *most R code is written by users*. In this chapter, you're going to learn how to create your own packages to share your code with your colleagues, friends, and the wider world. Even if you're a lone-working hermit who doesn't like sharing, packages are a great way to organize code for your own use.

Chapter Goals

After reading this chapter, you should:

- Be able to create a package
- Know how to document its functions and datasets
- Be able to release the package to CRAN

Why Create Packages?

The natural way to share R code and make it reusable by others (or even just yourself) is to package it up. In my experience, a lot of R users delay learning about how to create their own packages, perceiving it to be an advanced topic. In reality, it's a simple task—as long as you follow the prescribed rules. These rules are laid out in the "Writing R Extensions" manual that ships with R. If things go wrong, the answer is invariably buried within that document.

Prerequisites

Building packages requires a bunch of tools that are standard under Linux and other Unix derivatives, but not on Windows. All the tools have been collected together in a single download, available at *http://cran.r-project.org/bin/windows/Rtools* (or the *bin/windows/Rtools* directory of your nearest CRAN mirror). For even easier installation, use install.Rtools in the installr package.

While you're installing things, you'll want the devtools and roxygen2 packages as well:

```
install.packages(c("devtools", "roxygen2"))
```

The Package Directory Structure

Creating a package is mostly just a case of putting the right files in the right places. Inside your package directory, there are two compulsory files:

1. *DESCRIPTION* contains details about the package's version, its author, and what it does.

2. *NAMESPACE* describes which functions will be exported (made available to users).

Three other files are optionally allowed:

1. *LICENSE* (or *LICENCE*, depending upon which side of The Pond you hail from) contains the package license.

2. *NEWS* contains details on those exciting changes that you've made to the package.

3. *INDEX* contains names and descriptions for all the interesting objects in the package.

Before you start panicking at the thought of having to write five whole administrative files, take a deep breath. *NAMESPACE* and *INDEX* are completely autogenerated, *DESCRIPTION* is partially autogenerated, and you don't need a license file if you use one of several common, standard, licenses.[1]

At the top level, there are two directories that must be included:

1. *R* contains your R code.

2. *man* contains help files.

There are also some optional directories:

1. *src* contains C, C++, or Fortran source code.

2. *demo* contains R code demos to be run by the demo function.

3. *vignettes* contains longer documents explaining how to use the package, as found via browseVignettes.

---

1. *NEWS* is a pain in the ass, and you'll inevitably forget to update it.

---

4. *doc* contains help documents in other formats.

5. *data* contains data files.

6. *inst* contains anything else.

The first optional directories are beyond the scope of this quick introduction to package creation. Of the three compiled languages, C++ is easiest to use with R, thanks to the Rcpp package (see Dirk Eddelbuettel's *Seamless R and C++ Integration with Rcpp*). Creating vignettes isn't hard, especially if you use the knitr package (read Yihui Xie's *Dynamic Documents with R and knitr*).

Data files are things that will be made available via the data function (as we saw in "Built-in Datasets" on page 169 in Chapter 12). The preferred format is for them to be *.RData* files—the result of a call to save—though other formats are possible.

Although *inst* is the free-range folder that can contain anything, there are some standard contents that may be included:

1. *inst/tests* contains your RUnit or testthat tests.

2. *inst/python*, *inst/matlab*, and *inst/someotherscriptinglanguage* contain code from other scripting languages. (The three compiled languages that are supported go in *src*, as discussed above.)

3. *CITATION* describes how you would like the package to be cited, although this information is usually autogenerated from the description file.

Your First Package

OK, enough theory—let's make a package already. First, we need some contents: the hypotenuse function from the last chapter will do nicely. To demonstrate including data in a package, we can use some Pythagorean triples:

```
hypotenuse <- function(x, y)
{
  sqrt(x ^ 2 + y ^ 2)
}
pythagorean_triples <- data.frame(
  x = c(3, 5, 8, 7, 9, 11, 12, 13, 15, 16, 17, 19),
  y = c(4, 12, 15, 24, 40, 60, 35, 84, 112, 63, 144, 180),
  z = c(5, 13, 17, 25, 41, 61, 37, 85, 113, 65, 145, 181)
)
```

So now we need to create a load of directories to try and remember where to put things, right? Actually, it's easier than that. The package.skeleton function creates (almost) everything we need. It needs a name for the package ("pythagorus" will do nicely), and a character vector naming the variables to add to it:

```
package.skeleton(
  "pythagorus",
  c("hypotenuse", "pythagorean_triples")
)
```

Running `package.skeleton` creates the *R, man,* and *data* directories and the *DE-SCRIPTION* and *NAMESPACE* files, as well as a file named *Read-and-delete-me* that contains further instructions. Its output can be seen in Figure 17-1.

```
> package.skeleton(
+    "pythagorus",
+    c("hypotenuse", "pythagorean_triples")
+ )
Creating directories ...
Creating DESCRIPTION ...
Creating NAMESPACE ...
Creating Read-and-delete-me ...
Saving functions and data ...
Making help files ...
Done.
Further steps are described in './pythagorus/Read-and-delete-me'.
> |
```

Figure 17-1. Using package.skeleton to create the pythagorus package

The *DESCRIPTION* file has a strict structure of `Name: value` pairs. All we need to do is to update the `Title`, `Author`, `Maintainer`, `Description`, and `License` fields to something appropriate. The basic file format, created by `package.skeleton`, is shown in Figure 17-2.

```
DESCRIPTION
1 Package: pythagorus
2 Type: Package
3 Title: What the package does (short line)
4 Version: 1.0
5 Date: 2013-05-13
6 Author: Who wrote it
7 Maintainer: Who to complain to <yourfault@somewhere.net>
8 Description: More about what it does (maybe more than one line)
9 License: What license is it under?
10
```

Figure 17-2. The DESCRIPTION file created by package.skeleton

The `License` field must be one of "file" (in which case a *LICENCE* or *LICENSE* file must be included), "Unlimited" (no restriction), or one of these standard licenses: "GPL-2," "GPL-3," "LGPL-2," "LGPL-2.1," "LGPL-3," "AGPL-3," "Artistic-2.0," "BSD_2_clause," "BSD_3_clause," or "MIT."

The *NAMESPACE* file contains the text `exportPattern("^[[:alpha:]]+")`. This means "make any variable whose name begins with a letter available to the users."

Modern best practice is to write an `export` statement for each variable that you want to make available, rather than specifying a pattern.

We'll see how to automate creation of *NAMESPACE* in the next section, but for now, replace the text with `export(hypotenuse)`.

The *man* directory contains some automatically generated *.Rd* files: one for each function, one for each dataset, and another named *pythagorus-package.Rd*. These *.Rd* files contain LaTeX markup that will be used to create the help pages once the package is built. *pythagorus-package.Rd* contains a general help page for the whole package, where you can introduce the rest of the package. An example of an autogenerated *.Rd* file is given in Figure 17-3.

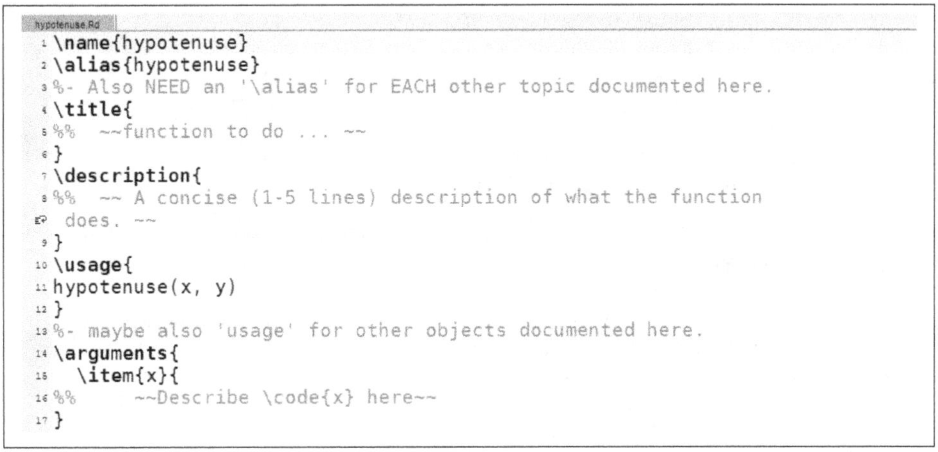

```
hypotenuse.Rd
1 \name{hypotenuse}
2 \alias{hypotenuse}
3 %- Also NEED an '\alias' for EACH other topic documented here.
4 \title{
5 %%   ~~function to do ... ~~
6 }
7 \description{
8 %%   ~~ A concise (1-5 lines) description of what the function
9 does. ~~
9 }
10 \usage{
11 hypotenuse(x, y)
12 }
13 %- maybe also 'usage' for other objects documented here.
14 \arguments{
15   \item{x}{
16 %%      ~~Describe \code{x} here~~
17 }
```

Figure 17-3. Autogenerated help page source file for the hypotenuse function

While the basic format is automatically created for you, R does not know what the function is for, so you need to manually fill in some of the details. There is a hard way to do this and an easy way, the latter of which is described next.

Documenting Packages

The big problem with having the help page for a function stuck in a different directory than the actual content is that it is very easy for the content to get out of sync. A typical example of this is when you add, remove, or rename the arguments that go into a function. R is not able to automatically change the corresponding help file to match the function, and you have to flit backward and forward between the two files to make sure that everything stays up to date.

The `roxygen2` package solves this problem by letting you write help text in comments next to the R code that you are trying to document. As a bonus, it uses a simple markup that reduces the amount of LaTeX that you need to know. roxygen2 is derived from Doxygen, which provides a similar documentation generation facility for C++, C, Java, Fortran, Python, and other languages. This makes it worthwhile learning the syntax, since you can then document code in many languages.

Each line of roxygen2 markup starts with #'. Some sections, like the title and description, are marked by their position at the start of the block. Other sections are denoted with a keyword. For example, the section describing the return values begins with @return. A full help block for a function would look like this:

```
#' Help page title
#'
#' A couple of lines of description about the function(s).
#' If you want to include code, use \code{my_code()}.
#' @param x Description of the first argument.
#' @param y Description of the second argument.
#' @return Description of the return value from a function.
#' If it returns a list, use
#' \itemize{
#'     \item{item1}{A description of item1.}
#'     \item{item2}{A description of item2.}
#' }
#' @note Describe how the algorithm works, or if the function has
#' any quirks here.
#' @author Your name here!
#' @references Journal papers, algorithms, or other inspiration here.
#' You can include web links like this
#' \url{http://www.thewebsiteyouarelinkingto.com}
#' @seealso Link to functions in the same package with
#' \code{\link{a_function_or_dataset}}
#' and functions in other packages with
#' \code{\link[another_package]{a_function_or_dataset}}
#' @examples
#' #R code run by the example function
#' \dontrun{
#' #R code that isn't run by example or when the package is built
#' }
#' @keywords misc
#' @export
f <- function(x, y)
{
    #Function content goes here, as usual
}
```

In the preceding example, there are a few things to pay special attention to.

The arguments are announced with the @param keyword. (The term "param" is standard throughout the Doxygen variants, so changing it to "arg" for R would cause more confusion than it would remove.) After @param comes a space, the name of the argument, and another space before the description of that argument.

Anything in the examples should be legal R code, since it is automatically run when you build the package. If you want to add commentary, use an extra hash (on top of the existing #' for roxygen2) to create R comments. If you want to add examples that may fail (demonstrating errors or creating files, for example), then wrap them in a \dontrun{} block.

Help files can contain keywords, but not just anything. To see the list of possible values, install the R.oo package, and run this snippet:

```
library(R.oo)
Rdoc$getKeywords()
```

(Alternatively, open the *KEYWORDS* file in the directory returned by R.home("doc").)

Adding the @export keyword lists the function in the *NAMESPACE* file, which in turn means that users should be able to call that function from the package, rather than it being an internal helper function.

The documentation for the whole package belongs in a file named *packagename-package.R*. It is similar to function documentation, but possibly even easier to write since there is less of it:

```
#' Help page title.  Probably the package name and tagline.
#'
#' A description of what the package does, why you might want to use it,
#' which functions to look at first, and anything else that the user
#' really, absolutely, must look at because you've created it and it is
#' astonishing.
#'
#' @author You again!
#' @docType package
#' @name packagename
#' @aliases packagename packagename-package
#' @keywords package
NULL
```

The two really important bits of function documentation are the @docType package line, which tells roxygen2 that this is whole-package documentation, and the NULL value afterward. This is needed for technical reasons—errors will result if you omit it.

Documenting datasets is almost the same as documenting the whole package. There is no standard place for this documentation; you can either append it to the package documentation file, or create a separate *packagename-data.R* file:

```
#' Help page title
#'
#' Explain the contents of each column here in the description.
#' \itemize{
#'   \item{column1}{Description of column1.}
#'   \item{column2}{Description of column2.}
#' }
```

```
#'
#' @references Where you found the data.
#' @docType data
#' @keywords datasets
#' @name datasetname
#' @usage data(datasetname)
#' @format A data frame with m rows of n variables
NULL
```

As with packages, the two important bits are the @docType data line, which tells roxygen2 that this is function documentation, and the NULL value afterward.

Once you've written documentation for each function, each dataset, and the whole package, call the roxygenize function to generate your help files and update the *NAMESPACE* and *DESCRIPTION* files (roxygenise is an identical alternative to roxygenize provided for those who prefer British spellings):

```
roxygenize("path/to/root/of/package")
```

Checking and Building Packages

Now you've created all the required directories, added R code and datasets, and documented them. You're nearly ready to build your package—the last task is to check that everything works.[2]

R has a built-in check tool, R CMD check, available from your OS command line. It's incredibly thorough and is the main reason that most packages you download from CRAN actually work. Of course, using a DOS or bash command line is so 20th century —a better alternative is to use the check function in the devtools package, the output of which is shown in Figure 17-4:

```
library(devtools)
check("path/to/root/of/package")
```

This provides several pages of output and will warn on a variety of things, like documentation not matching the corresponding function, names not being portable across OSs, examples not running correctly, and whether it thought your coding style was a bit old-fashioned. (OK, the last one is made up, but there are lots of checks.)

---

2. Spoiler: the answer is usually, "No, you forgot something."

```
> check("pythagorus")
Updating pythagorus documentation
Loading pythagorus
Writing pythagorean_triples.Rd
"C:/PROGRA~1/R/R-devel/bin/x64/R" --vanilla CMD build  \
  "d:\workspace\pythagorus" --no-manual --no-resave-data

* checking for file 'd:\workspace\pythagorus/DESCRIPTION' ... OK
* preparing 'pythagorus':
* checking DESCRIPTION meta-information ... OK
* checking for LF line-endings in source and make files
* checking for empty or unneeded directories
Removed empty directory 'pythagorus/inst'
* looking to see if a 'data/datalist' file should be added
* building 'pythagorus_1.0.tar.gz'

"C:/PROGRA~1/R/R-devel/bin/x64/R" --vanilla CMD check  \
```

Figure 17-4. Output when checking a package

Read the output, fix the errors and warnings, and repeat. Once you are satisfied that your package files are mistake-free, you can finally build it! As with check, there is a command-line version of build built into R, but it is much easier to use the function in the devtools package. You get a choice of building to source (portable across OSs; the standard format for Linux) or binary (specific to your current OS):

```
build("path/to/root/of/package")
```

That's it—you now have your own package! Although, wouldn't it be great if *other people* had your package too? To release your package to CRAN, use the devtools release function:

```
release("path/to/root/of/package")
```

This asks you lots of questions to make sure that you're really, really, really sure you've done everything you need to before you send your package.

 When you upload a package to CRAN, R Core members will check that your package builds without warnings. Their time is precious, so it's important that you run the check function and that you fix *all* the errors and warnings before uploading to CRAN.

Maintaining Packages

Functions act like a black box to users. They pass in some arguments and a value gets returned, without the user needing to be aware of what happens inside the function (at least in theory). This means that the signature of a function (the names and order of the function's arguments) should not be changed without warning the user. R provides several functions to help notify users of changes to signatures.

If you are planning to add a new feature but haven't had the chance to implement it, or you want to give your users advance notice that this feature is coming, use the `.NotYetUsed` function. This causes an error or a warning that the argument is not yet in use, which will be thrown if a user tries to use it prematurely. In this next example, we're going to extend the hypotenuse to work as a two-dimensional p-norm. Before we add the new functionality, we're just going to change the signature, and throw an error if the user tries to use the p argument:

```
hypotenuse <- function(x, y, p = 2)
{
  if(!missing(p))
  {
    .NotYetUsed("p")
  }
  sqrt(x ^ 2 + y ^ 2)
}
hypotenuse(5, 12)        #behavior as before

## [1] 13

hypotenuse(5, 12, 1)

## Error: argument 'p' is not used (yet)
```

Once we add the new functionality, we can remove the call to `.NotYetUsed`:

```
hypotenuse <- function(x, y, p = 2)
{
  (x ^ p + y ^ p) ^ (1 / p)
}
```

If you want to add a whole new function (rather than just an argument), the equivalent function is `.NotYetImplemented`. This is brilliant when you are first creating a package, or adding a large block of functionality. Writing a single function can be time-consuming, so by the time you've written a few you may have forgotten which others you were going to add. Sometimes it's better to work at a high level and fill in the details later. Simply create a placeholder for each function, making the body a call to `.NotYetImplemented`. One day the function in the next example will calculate triangular numbers, but for now it just throws an error:

```
triangular <- function(n)
{
  .NotYetImplemented()
}
triangular()

## Error: 'triangular' is not implemented yet
```

If you want to remove a function, it is polite to do it in stages. The first step is to add a call to .Deprecated, with the name of an alternative function to call instead. The rest of the function should remain unchanged, so that the existing behavior is preserved:

```
hypotenuse <- function(x, y, p = 2)
{
  .Deprecated("p_norm")
  (x ^ p + y ^ p) ^ (1 / p)
}
hypotenuse(5, 12)

## Warning: 'hypotenuse' is deprecated. Use 'p_norm' instead. See
## help("Deprecated")

## [1] 13
```

After a suitable length of time—long enough for your users to have noticed the messages about deprecation—you can change the contents of the function to a call to .Defunct, which throws an error:

```
hypotenuse <- function(x, y, p = 2)
{
  .Defunct("p_norm")
}
hypotenuse(5, 12)

## Error: 'hypotenuse' is defunct. Use 'p_norm' instead. See help("Defunct")
```

Summary

- Making packages mostly involves organizing your files into a specific structure.
- package.skeleton creates much of this structure for you.
- The roxygen2 package makes documenting packages easy.
- The devtools package makes checking and building packages easy.
- NotYetImplemented, Deprecated, and Defunct help you maintain your packages.

Test Your Knowledge: Quiz

Question 17-1
Which of the five files in the top-level directory of an R package are compulsory? The files are *DESCRIPTION, INDEX, LICENSE, NAMESPACE,* and *NEWS.*

Question 17-2
Which of the eight directories in an R package are compulsory? The directories are *data, demo, doc, inst, man, R, src,* and *vignettes.*

Question 17-3

Why might you include a *CITATION* file in your package?

Question 17-4

Which function must you call to generate help files from roxygen2 markup?

Question 17-5

What's the recommended way to politely remove a function from a package?

Test Your Knowledge: Exercises

Exercise 17-1

1. Write a function, sum_of_squares, that calculates the sum of the first *n* square numbers. (Hint: the formula is *n* * (*n* + 1) * (2 * *n* + 1) / 6.) [5]

2. Create a data frame, squares_data, with two columns. The x column should contain the numbers 1 to 10. The y column should contain the sum of the square numbers up to the corresponding x value—that is, the result of sum_of_squares(1:10). [5]

3. Use package.skeleton to create the squares package, containing the sum_of_squares function and the squares_data data frame. [5]

Exercise 17-2

Document the sum_of_squares function, the squares_data data frame, and the whole squares package using roxygen2 markup, and generate the help files.

Exercise 17-3

Use the devtools package to check and build the squares package.

## PART III

Appendixes

PART IV

Appendixes

Properties of Variables

R has four different functions that tell you the type of a variable, namely `class`, `typeof`, `mode`, and `storage.mode`. For some variable types they all give the same answer, but for others they differ, and it gets a little bit complicated.

For the vast majority of code you write, you'll only ever care about the `class`. The only time `class` can't help you is when you are examining matrices or arrays, and you care about whether the variable contains numbers or characters (or some other type). In this case, you can use one of the other three type functions, or call one of the `is.*` functions (`is.numeric`, for example).

Table A-1 shows the values returned by `class`, `typeof`, `mode`, and `storage.mode` for various variable types.

Table A-1. Comparison of variable class, type, mode, and storage mode

	class	typeof	mode	storage.mode
Logical	logical	logical	logical	logical
Integer	integer	integer	numeric	integer
Floating Point	numeric	double	numeric	double
Complex	complex	complex	complex	complex
String	character	character	character	character
Raw byte	raw	raw	raw	raw
Categorical	factor	integer	numeric	integer
Null	NULL	NULL	NULL	NULL
Logical Matrix	matrix	logical	logical	logical
Numeric Matrix	matrix	double	numeric	double
Character Matrix	matrix	character	character	character
Logical Array	array	logical	logical	logical

	class	typeof	mode	storage.mode
Numeric Array	array	double	numeric	double
Character Array	array	character	character	character
List	list	list	list	list
Data Frame	data.frame	list	list	list
Function	function	closure	function	function
Environment	environment	environment	environment	environment
Expression	expression	expression	expression	expression
Call	call	language	call	language
Formula	formula	language	call	language

In R, vectors are variable types with a length, but no dimension (that is, dim returns NULL) and no attributes other than names. Vector types include numeric, logical, and character types, but also lists and expressions. The rule about no attributes means that factors are not vectors.

 Lists *are* vectors. Factors are *not* vectors.

Related to vectors are atomic types. *Atomic* means that a type cannot contain other instances of that type within itself. The opposite of atomic is *recursive*: lists are the canonical example, since they can contain other lists. An object can only ever be atomic or recursive, never both.

 Matrices and arrays are atomic.

Some objects are known as *language objects*. These variable types can be evaluated to run R code.

Table A-2 shows the values returned by is.vector, is.atomic, is.recursive, and is.language for various variable types.

Table A-2. Comparison of which variable types are vectors, atomic, recursive, or language objects

	is.vector	is.atomic	is.recursive	is.language
Logical	TRUE	TRUE	FALSE	FALSE
Integer	TRUE	TRUE	FALSE	FALSE
Floating Point	TRUE	TRUE	FALSE	FALSE
Complex	TRUE	TRUE	FALSE	FALSE
String	TRUE	TRUE	FALSE	FALSE
Raw Byte	TRUE	TRUE	FALSE	FALSE
Categorical	FALSE	TRUE	FALSE	FALSE
Null	FALSE	TRUE	FALSE	FALSE
Logical Matrix	FALSE	TRUE	FALSE	FALSE
Numeric Matrix	FALSE	TRUE	FALSE	FALSE
Character Matrix	FALSE	TRUE	FALSE	FALSE
Logical Array	FALSE	TRUE	FALSE	FALSE
Numeric Array	FALSE	TRUE	FALSE	FALSE
Character Array	FALSE	TRUE	FALSE	FALSE
List	TRUE	FALSE	TRUE	FALSE
Data Frame	FALSE	FALSE	TRUE	FALSE
Function	FALSE	FALSE	TRUE	FALSE
Environment	FALSE	FALSE	TRUE	FALSE
Expression	TRUE	FALSE	TRUE	TRUE
Call	FALSE	FALSE	TRUE	TRUE
Formula	FALSE	FALSE	TRUE	TRUE

There are lots of things that R can do that didn't make it into this book, either because they are niche requirements, because they're too advanced, or because there are really good books out there already dealing with those topics. This appendix is intended to point you in the direction of further resources. Several of the subjects covered here are dealt with in more detail in the CRAN Task Views (*http://cran.r-project.org/web/views/*), which are well worth browsing.

You can make GUIs for your R code to help less technical users. The gWidgets framework allows high-level access to several graphics toolkits, including tcl/tk, qt, GTK2, and HTML. (Low-level access is also possible via tcltk/tcltk2, qtbase, RGtk2, and several HTML-generation packages.) Read *Programming Graphical User Interfaces in R* by Michael Lawrence and John Verzani for more information. The Java-based Deducer package provides an alternative, and RStudio's shiny package makes it simple to make web apps with R.

You can call code in compiled languages (C, C++, and FORTRAN) via the .Call function. By far the easiest way of using these languages is to write C++ in conjunction with the Rcpp package. Read *Seamless R and C++ Integration with Rcpp* by Dirk Eddelbuettel for more on this subject.

R is single threaded, but there are several ways of making it use multiple cores or multiple machines to parallelize your code. The parallel package ships with R, and under POSIX-compliant operating systems (that's most OSs that aren't Windows) you can simply replace calls to lapply with mclapply to run loops over all the cores on your machine. Multicore capabilities using (MPI, SOCK, or other) socket-based clusters are possible using the snow and parallel packages. You can also connect to a Hadoop cluster using the RHIPE (*http://www.datadr.org*) and rmr packages, or Hadoop in an Amazon Web Services using segue (*http://code.google.com/p/segue/*), not available for

Windows. Read *Parallel R* by Q. Ethan McCallum and Stephen Weston for more details. The Enterprise edition of Revolution R has some parallelization features built into it.

The `ff` and `bigmemory` packages allow you to store R variables in files (like SAS does), avoiding the limitations of RAM. On a related note, the `data.table` package provides an enhanced data frame variable type that has faster indexing, assignment, and merging capabilities.

There are dozens of spatial statistics packages: `sp` provides a standard way of storing spatial data objects; `maps`, `maptools`, and `mapproj` provide helper functions for reading and writing maps; `spatstat` provides functions for (you guessed it) spatial statistics; `OpenStreetMap` retrieves raster images from *http://openstreetmap.com*; and so on.

Finally, R has a tool for combining code and output with regular text in a report (sometimes called *literate programming* or *reproducible research*), called `Sweave`. This has been improved upon and extended by the `knitr` package, which allows you to create reports using a variety of markup languages. In fact, this book was written using `knitr` to create AsciiDoc markup, which in turn was used to create PDF, HTML, and ebook documents. Yihui Xie's *Dynamic Documents with R and knitr* explains how to use it.

The R ecosystem is big, and getting bigger all the time. These are just a few of the things that you might want to explore; take a look in the bibliography for some more ideas. Have fun exploring!

Answers to Quizzes

Question 1-1

R is an open source reworking of the S programming language.

Question 1-2

Choices include imperative, object-oriented, and functional.

Question 1-3

```
8:27
```

Question 1-4

`help.search` (which does the same as `??`)

Question 1-5

`RSiteSearch`

Question 2-1

```
%/%
```

Question 2-2

`all.equal(x, pi)` or, even better, `isTRUE(all.equal(x, pi))`

Question 2-3

At least two of the following:

1. `<-`
2. `+ =`
3. `+ <<-`
4. `assign`

Question 2-4

Just `Inf` and `-Inf`

Question 2-5

0, Inf, and -Inf

Question 3-1

numeric, integer, and complex

Question 3-2

nlevels

Question 3-3

as.numeric("6.283185")

Question 3-4

Any three of summary, head, str, unclass, attributes, or View. Bonus points if you've discovered tail, the counterpart to head that prints the last few rows.

Question 3-5

rm(list = ls())

Question 4-1

seq.int(0, 1, 0.25)

Question 4-2

Either by using *name* = *value* pairs when the vector is created, or by calling the names function afterward.

Question 4-3

Positive integers for locations to retrieve, negative integers for locations to avoid, logical values, or names of elements.

Question 4-4

3 * 4 * 5 = 60

Question 4-5

%*%

Question 5-1

3. The inner list counts as one element, and so does the NULL element.

Question 5-2

When passing arguments to functions, when calling formals, or in the global environment variable .Options.

Question 5-3

You can use matrix-style indexing with pairs of positive integers/negative integers/ logical values/characters in single square brackets. You can also use list-style indexing with one index value inside single or double square brackets, or the dollar sign ($) operator. Thirdly, you can call the subset function.

Question 5-4

By passing check.names = FALSE to data.frame.

Question 5-5

rbind for appending vertically or cbind for appending horizontally.

Question 6-1

The user workspace.

Question 6-2

list2env is the best solution, but as.environment also works.

Question 6-3

Just type its name.

Question 6-4

formals, args, and formalArgs.

Question 6-5

do.call calls a function with its arguments in a list form.

Question 7-1

format, formatC, sprintf, and prettyNum are the main ones.

Question 7-2

Using alarm, or printing an \a character to the console.

Question 7-3

factor and ordered

Question 7-4

The value is counted as missing (NA).

Question 7-5

Use cut to bin it.

Question 8-1

if will throw an error if you pass NA to it.

Question 8-2

ifelse will return NA values in the corresponding positions where NA is passed to it.

Question 8-3

switch will conditionally execute code based upon a character or integer argument.

Question 8-4

Insert the keyword `break` into your loop code.

Question 8-5

Insert the keyword `next` into your loop code.

Question 9-1

`lapply`, `vapply`, `sapply`, `apply`, `mapply`, and `tapply` were all discussed in the chapter, with `eapply` and `rapply` getting brief mentions too. Try `apropos("apply")` to see all of them.

Question 9-2

All three functions accept a list and apply a function to each element in turn. The difference is in the return value. `lapply` always returns a list, `vapply` always returns a vector or array as specified by a template, and `sapply` can return either.

Question 9-3

`rapply` is recursive, and ideal for deeply nested objects like trees.

Question 9-4

This is a classic split-apply-combine problem. Use `tapply` (or something from the `plyr` package).

Question 9-5

In a name like `**ply`, the first asterisk denotes the type of the first input argument and the second asterisk denotes the type of the return value.

Question 10-1

CRAN is by far the biggest package repository. Bioconductor, R-Forge, and RForge.net are others. There are also many packages on GitHub, Bitbucket, and Google Code.

Question 10-2

Both functions load a package, but `library` throws an error if it fails, whereas `require` returns a logical value (letting you do custom error handling).

Question 10-3

A package library is just a folder on your machine that contains R packages.

Question 10-4

`.libPaths` returns a list of libraries.

Question 10-5

R doesn't do a great impression of Internet Explorer, but you can make it use Internet Explorer's *internet2.dll* library for connecting to the Internet.

Question 11-1

POSIXct classes must be used. Dates don't store the time information, and POS IXlt dates store their data as lists, which won't fit inside a data frame.

Question 11-2

Midnight at the start of January 1, 1970.

Question 11-3

"%B %Y"

Question 11-4

Add 3,600 seconds to it. For example:

```
x <- Sys.time()
x + 3600

## [1] "2013-07-17 22:44:55 BST"
```

Question 11-5

The period will be longer, because 2016 is a leap year. A duration is always exactly 60 * 60 * 24 * 365 seconds. A period of one year will be 366 days in a leap year.

Question 12-1

Call the data function with no arguments.

Question 12-2

read.csv assumes that a decimal place is represented by a full stop (period) and that each item is separated by a comma, whereas read.csv2 assumes that a decimal place is represented by a comma and that each item is separated by a semicolon. read.csv is used for data created in locales where a period is used as a decimal place (most English-speaking locales, for example). read.csv2 is for data created in locales where a comma is used (most European locales, for example). If you are unsure, simply open your data file in a text editor.

Question 12-3

read.xlsx2 from the xlsx package is a good first choice, but there is also read.xlsx in the same package, and different functions in several other packages.

Question 12-4

You can simply pass the URL to read.csv, or use download.file to get a local copy.

Question 12-5

Currently SQLite, MySQL, PostgreSQL, and Oracle databases are supported.

Question 13-1

Read in the text as a character vector wih readLines, call str_count to count the number of instances in each line, and sum the total.

Question 13-2

`with`, `within`, `transform`, and `mutate` all allow manipulating columns and adding columns to data frames, as well as standard assignment.

Question 13-3

Casting. Not *freezing*!

Question 13-4

Use `order` or `arrange`.

Question 13-5

Define a function that reads `TRUE` when you have a positive number—for example, `is.positive <- function(x) x > 0`—and call `Find(is.positive, x)`.

Question 14-1

`min` returns the single smallest value of all its inputs. `pmin` accepts several vectors that are the same length, and returns the smallest at each point along them.

Question 14-2

Pass the `pch` ("plot character") argument.

Question 14-3

Use a formula of the form `y ~ x`.

Question 14-4

An aesthetic specifies a variable that you will look at a variation in. Most plots take an *x* and a *y* aesthetic for x and y coordinates, respectively. You can also specify color aesthetics or shape aesthetics (where more than two variables are to be looked at at once, for example).

Question 14-5

Histograms, box plots, and kernel density plots were all mentioned in the chapter. There are some other weirdly esoteric plots, like violin plots, rug plots, bean plots, and stem-and-leaf plots, that weren't mentioned. Have 100 geek points for each of these that you guessed.

Question 15-1

Set the seed (with `set.seed`), generate the numbers, then reset the seed to the same value.

Question 15-2

PDF functions have a name beginning with d, followed by the name of the distribution. For example, the PDF for the binomial distribution is `dbinom`. CDF functions start with p followed by the name of the distribution, and inverse CDF functions begin with q followed by the name of the distribution.

Question 15-3

Colons represent an interaction between variables.

Question 15-4

anova, AIC, and BIC are common functions for comparing models.

Question 15-5

The R^2 value is available via summary(model)$r.squared.

Question 16-1

The warnings function shows the previous warnings.

Question 16-2

Upon failure, try returns an object of class try-error.

Question 16-3

The testthat equivalent of checkException is expect_exception. Say that 10 times fast.

Question 16-4

quote turns a string into a call, then eval evaluates it.

Question 16-5

Overload functions using the S3 system. A function print.foo will be called for objects of class foo.

Question 17-1

DESCRIPTION and *NAMESPACE* are compulsory.

Question 17-2

man and *R* are compulsory in all packages. *src* is required if you include C, C++, or Fortran code.

Question 17-3

CITATION files let you explain who made and maintains the package, if that information is too long and complicated to go in the *DESCRIPTION* file.

Question 17-4

roxygenise or roxygenize

Question 17-5

First warn the user that it is deprecated by adding a call to .Deprecated. Later, replace the body completely with a call to .Defunct.

Solutions to Exercises

Exercise 1-1

If you get stuck, ask your system administrator, or ask on the *R-help* mailing list.

Exercise 1-2

Use the colon operator to create a vector, and the sd function:

```
sd(0:100)
```

Exercise 1-3

Type demo(plotmath) and hit Enter, or click the plots to see what's on offer.

Exercise 2-1

1. Simple division gets the reciprocal, and atan calculates the inverse (arc) tangent:

   ```
   atan(1 / 1:1000)
   ```

2. Assign variables using <-:

   ```
   x <- 1:1000
   y <- atan(1 / x)
   z <- 1 / tan(y)
   ```

Exercise 2-2

For comparing two vectors that should contain the same numbers, all.equal is usually what you need:

```
x == z
identical(x, z)
all.equal(x, z)
all.equal(x, z, tolerance = 0)
```

Exercise 2-3

The exact values contained in the following three vectors may vary:

```
true_and_missing <- c(NA, TRUE, NA)
false_and_missing <- c(FALSE, FALSE, NA)
mixed <- c(TRUE, FALSE, NA)

any(true_and_missing)
any(false_and_missing)
any(mixed)
all(true_and_missing)
all(false_and_missing)
all(mixed)
```

Exercise 3-1

```
class(Inf)
class(NA)
class(NaN)
class("")
```

Repeat with typeof, mode, and storage.mode.

Exercise 3-2

```
pets <- factor(sample(
  c("dog", "cat", "hamster", "goldfish"),
  1000,
  replace = TRUE
))
head(pets)
summary(pets)
```

Converting to factors is recommended but not compulsory.

Exercise 3-3

```
carrot <- 1
potato <- 2
swede  <- 3
ls(pattern = "a")
```

Your vegetables may vary.

Exercise 4-1

There are several possibilities for creating sequences, including the colon operator. This solution uses seq_len and seq_along:

```
n <- seq_len(20)
triangular <- n * (n + 1) / 2
names(triangular) <- letters[seq_along(n)]
triangular[c("a", "e", "i", "o")]
```

Exercise 4-2

Again, there are many different ways of creating a sequence from -11 to 0 to 11. abs gives you the absolute value of a number:

```
diag(abs(seq.int(-11, 11)))
```

Exercise 4-3

Wilkinson matrices have the interesting property that most of their eigenvalues form pairs of nearly equal numbers. The 21-by-21 matrix is the most frequently used:

```
identity_20_by_21 <- diag(rep.int(1, 20), 20, 21)
below_the_diagonal <- rbind(0, identity_20_by_21)
identity_21_by_20 <- diag(rep.int(1, 20), 21, 20)
above_the_diagonal <- cbind(0, identity_21_by_20)
on_the_diagonal <- diag(abs(seq.int(-10, 10)))
wilkinson_21 <- below_the_diagonal + above_the_diagonal + on_the_diagonal
eigen(wilkinson_21)$values
```

Exercise 5-1

For simplicity, here I've manually specified which numbers are square. Can you think of a way to automatically determine if a number is square?

```
list(
    "0 to 9"   = c(0, 1, 4, 9),
    "10 to 19" = 16,
    "20 to 29" = 25,
    "30 to 39" = 36,
    "40 to 49" = 49,
    "50 to 59" = NULL,
    "60 to 69" = 64,
    "70 to 79" = NULL,
    "80 to 89" = 81,
    "90 to 99" = NULL
)
```

More fancily, we can automate the calculation for square numbers. Here, the cut function creates different groups with the ranges 0 to 9, 10 to 19, and so on, then returns a vector stating which group each square number is in. Then the split function splits the square number vector into a list, with each element containing the values from the corresponding group:

```
x <- 0:99
sqrt_x <- sqrt(x)
is_square_number <- sqrt_x == floor(sqrt_x)
square_numbers <- x[is_square_number]
groups <- cut(
    square_numbers,
    seq.int(min(x), max(x), 10),
    include.lowest = TRUE,
    right = FALSE
)
split(square_numbers, groups)
```

Exercise 5-2

There are many different ways of getting the numeric subset of the `iris` dataset. Experiment with them!

```
iris_numeric <- iris[, 1:4]
colMeans(iris_numeric)
```

Exercise 5-3

This solution is just one of the many ways of indexing and subsetting the data frames:

```
beaver1$id <- 1
beaver2$id <- 2
both_beavers <- rbind(beaver1, beaver2)
subset(both_beavers, as.logical(activ))
```

Exercise 6-1

We create an environment using `new.env`. After that, the syntax is the same as with lists:

```
multiples_of_pi <- new.env()
multiples_of_pi[["two_pi"]] <- 2 * pi
multiples_of_pi$three_pi <- 3 * pi
assign("four_pi", 4 * pi, multiples_of_pi)
ls(multiples_of_pi)
```

```
## [1] "four_pi"  "three_pi" "two_pi"
```

Exercise 6-2

This is easier than you think. We just use the modulo operator to get the remainder when divided by two, and everything automatically works, including nonfinite values:

```
is_even <- function(x) (x %% 2) == 0
is_even(c(-5:5, Inf, -Inf, NA, NaN))
```

```
##  [1] FALSE  TRUE FALSE  TRUE FALSE  TRUE FALSE  TRUE FALSE  TRUE FALSE
## [12]    NA    NA    NA    NA
```

Exercise 6-3

Again, the function is just a one-liner. The `formals` and body functions do the hard work, and we just need to return a list of their results:

```
args_and_body <- function(fn)
{
  list(arguments = formals(fn), body = body(fn))
}
args_and_body(var)
```

```
## $arguments
## $arguments$x
##
##
## $arguments$y
```

```
## NULL
##
## $arguments$na.rm
## [1] FALSE
##
## $arguments$use
##
##
##
## $body
## {
##     if (missing(use))
##         use <- if (na.rm)
##             "na.or.complete"
##         else "everything"
##     na.method <- pmatch(use, c("all.obs", "complete.obs",
##         "pairwise.complete.obs", "everything", "na.or.complete"))
##     if (is.na(na.method))
##         stop("invalid 'use' argument")
##     if (is.data.frame(x))
##         x <- as.matrix(x)
##     else stopifnot(is.atomic(x))
##     if (is.data.frame(y))
##         y <- as.matrix(y)
##     else stopifnot(is.atomic(y))
##     .Call(C_cov, x, y, na.method, FALSE)
## }

args_and_body(alarm)

## $arguments
## NULL
##
## $body
## {
##     cat("\a")
##     flush.console()
## }
```

Exercise 7-1

```
formatC(pi, digits = 16)

## [1] "3.141592653589793"
```

This also works if you replace formatC with format or prettyNum.

Exercise 7-2

Call strsplit with a regular expression to match the spaces and punctuation. Recall that ? makes the matching character optional:

```
#split on an optional comma, a compulsory space, an optional
#hyphen, and an optional space
strsplit(x, ",? -? ?")
```

```
## [[1]]
## [1] "Swan"   "swam"   "over"   "the"    "pond"  "Swim"  "swan"  "swim!"
##
## [[2]]
## [1] "Swan"   "swam"   "back"   "again" "Well"  "swum"  "swan!"

#or, grouping the final hyphen and space
strsplit(x, ",? (- )?")

## [[1]]
## [1] "Swan"   "swam"   "over"   "the"    "pond"  "Swim"  "swan"  "swim!"
##
## [[2]]
## [1] "Swan"   "swam"   "back"   "again" "Well"  "swum"  "swan!"
```

Exercise 7-3

By default, cut creates intervals that include an upper bound but not a lower bound. In order to get a count for the smallest value (3), we need an extra break point below it (at 2, for example). Try replacing table with count from the plyr package:

```
scores <- three_d6(1000)
bonuses <- cut(
  scores,
  c(2, 3, 5, 8, 12, 15, 17, 18),
  labels = -3:3
)
table(bonuses)

## bonuses
##  -3  -2  -1   0   1   2   3
##   4  39 186 486 233  47   5
```

Exercise 8-1

Use if to distinguish the conditions. The %in% operator makes checking the conditions easier:

```
score <- two_d6(1)
if(score %in% c(2, 3, 12))
{
   game_status <- FALSE
   point <- NA
} else if(score %in% c(7, 11))
{
   game_status <- TRUE
   point <- NA
} else
{
   game_status <- NA
   point <- score
}
```

Exercise 8-2

Since we want to execute the code an unknown number of times, the repeat loop is most appropriate:

```
if(is.na(game_status))
{
  repeat({
    score <- two_d6(1)
    if(score == 7)
    {
      game_status <- FALSE
      break
    } else
    if(score == point)
    {
      game_status <- TRUE
      break
    }
  })
}
```

Exercise 8-3

Since we know how many times we want to execute the loop (from the length of the shortest word to the length of the longest word), the for loop is most appropriate:

```
nchar_sea_shells <- nchar(sea_shells)

for(i in min(nchar_sea_shells):max(nchar_sea_shells))
{
  message("These words have ", i, " letters:")
  print(toString(unique(sea_shells[nchar_sea_shells == i])))
}
## These words have 2 letters:
## [1] "by, So, if, on"
## These words have 3 letters:
## [1] "She, sea, the, The, she, are, I'm"
## These words have 4 letters:
## [1] "sure"
## These words have 5 letters:
## [1] "sells"
## These words have 6 letters:
## [1] "shells, surely"
## These words have 7 letters:
## [1] ""
```

```
## These words have 8 letters:

## [1] "seashore"

## These words have 9 letters:

## [1] "seashells"
```

Exercise 9-1

Since the input is a list and the output is always the same known length, vapply is the best choice:

```
vapply(wayans, length, integer(1))
```

```
##   Dwayne Kim Keenen Ivory      Damon      Kim     Shawn
##              0         5          4        0         3
##       Marlon         Nadia     Elvira   Diedre    Vonnie
##            2             0          2        5         0
```

Exercise 9-2

1. We can get a feel for the dataset by using str, head, and class, and by reading the help page ?state.x77:

```
##  num [1:50, 1:8] 3615 365 2212 2110 21198 ...
##  - attr(*, "dimnames")=List of 2
##  ..$ : chr [1:50] "Alabama" "Alaska" "Arizona" "Arkansas" ...
##  ..$ : chr [1:8] "Population" "Income" "Illiteracy" "Life Exp" ...

##             Population Income Illiteracy Life Exp Murder HS Grad Frost
## Alabama          3615   3624        2.1    69.05   15.1    41.3    20
## Alaska            365   6315        1.5    69.31   11.3    66.7   152
## Arizona          2212   4530        1.8    70.55    7.8    58.1    15
## Arkansas         2110   3378        1.9    70.66   10.1    39.9    65
## California      21198   5114        1.1    71.71   10.3    62.6    20
## Colorado         2541   4884        0.7    72.06    6.8    63.9   166
##                 Area
## Alabama        50708
## Alaska        566432
## Arizona       113417
## Arkansas       51945
## California    156361
## Colorado      103766

## [1] "matrix"
```

2. Now that we know that the data is a matrix, the apply function is most appropriate. We want to loop over columns, so we pass 2 for the dimension value:

```
## Population      Income Illiteracy   Life Exp     Murder    HS Grad
##   4246.420    4435.800      1.170     70.879      7.378     53.108
##      Frost        Area
##    104.460   70735.880

## Population      Income Illiteracy   Life Exp     Murder    HS Grad
##  4.464e+03   6.145e+02  6.095e-01  1.342e+00  3.692e+00  8.077e+00
```

```
##      Frost      Area
## 5.198e+01  8.533e+04
```

Exercise 9-3

tapply is the best choice from base R. ddply from plyr also works nicely:

```
with(commute_data, tapply(time, mode, quantile, prob = 0.75))
```

```
## bike   bus   car train
## 63.55 55.62 42.33 36.29
```

```
ddply(commute_data, .(mode), summarize, time_p75 = quantile(time, 0.75))
```

```
##     mode time_p75
## 1  bike    63.55
## 2   bus    55.62
## 3   car    42.33
## 4 train    36.29
```

Exercise 10-1

In R GUI, click Packages → "Install package(s)...," choose a mirror, and then choose Hmisc. If the download or install fails, then make sure that you have Internet access and permission to write to the *library* directory. You can also try visiting the CRAN website and doing a manual download. The most common cause of problems is restricted permissions; speak to your sys admin if this is the case.

Exercise 10-2

```
install.packages("lubridate")
```

You might want to specify a library to install to.

Exercise 10-3

The installed.packages function retrieves the details of the packages on your machine and their locations (the results will be different for your machine):

```
pkgs <- installed.packages()
table(pkgs[, "LibPath"])
```

```
##
## C:/Program Files/R/R-devel/library                  D:/R/library
##                               29                             169
```

Exercise 11-1

This example uses strptime to parse and strftime to format, though there are many possibilities:

```
in_string <- c("1940-07-07", "1940-10-09", "1942-06-18", "1943-02-25")
(parsed <- strptime(in_string, "%Y-%m-%d"))
```

```
## [1] "1940-07-07" "1940-10-09" "1942-06-18" "1943-02-25"
```

```
(out_string <- strftime(parsed, "%a %d %b %y"))
```

```
## [1] "Sun 07 Jul 40" "Wed 09 Oct 40" "Thu 18 Jun 42" "Thu 25 Feb 43"
```

Exercise 11-2

The ?Sys.timezone help page suggests this code for importing the time zone file:

```
tzfile <- file.path(R.home("share"), "zoneinfo", "zone.tab")
tzones <- read.delim(
  tzfile,
  row.names = NULL,
  header = FALSE,
  col.names = c("country", "coords", "name", "comments"),
  as.is = TRUE,
  fill = TRUE,
  comment.char = "#"
)
```

The best way to find your own time zone depends on where you are. Start with View(tzones) and use subset to narrow your search if you need to.

Exercise 11-3

You can solve this using base R by accessing components of a POSIXlt date, but it's a little bit cleaner to use the lubridate month and day functions:

```
zodiac_sign <- function(x)
{
  month_x <- month(x, label = TRUE)
  day_x <- day(x)
  switch(
    month_x,
    Jan = if(day_x < 20) "Capricorn" else "Aquarius",
    Feb = if(day_x < 19) "Aquarius" else "Pisces",
    Mar = if(day_x < 21) "Pisces" else "Aries",
    Apr = if(day_x < 20) "Aries" else "Taurus",
    May = if(day_x < 21) "Taurus" else "Gemini",
    Jun = if(day_x < 21) "Gemini" else "Cancer",
    Jul = if(day_x < 23) "Cancer" else "Leo",
    Aug = if(day_x < 23) "Leo" else "Virgo",
    Sep = if(day_x < 23) "Virgo" else "Libra",
    Oct = if(day_x < 23) "Libra" else "Scorpio",
    Nov = if(day_x < 22) "Scorpio" else "Sagittarius",
    Dec = if(day_x < 22) "Sagittarius" else "Capricorn"
  )
}
#Usage is, for example,
nicolaus_copernicus_birth_date <- as.Date("1473-02-19")
zodiac_sign(nicolaus_copernicus_birth_date)

## [1] "Pisces"
```

Note that the switch statement means that this implementation of the function isn't vectorized. You can achieve vectorization by using lots of ifelse statements or, more easily, by calling Vectorize(zodiac_sign).

Exercise 12-1

Use `system.file` to locate the file and `read.csv` to import the data:

```
hafu_file <- system.file("extdata", "hafu.csv", package = "learningr")
hafu_data <- read.csv(hafu_file)
```

Exercise 12-2

Use `read.xlsx2` from the `xlsx` package to import the data. The data is in the first (and only) sheet, and it is best to specify the types of data in each column:

```
library(xlsx)
gonorrhoea_file <- system.file(
  "extdata",
  "multi-drug-resistant gonorrhoea infection.xls",
  package = "learningr"
)
gonorrhoea_data <- read.xlsx2(
  gonorrhoea_file,
  sheetIndex = 1,
  colClasses = c("integer", "character", "character", "numeric")
)
```

Exercise 12-3

There are several ways of doing this using the `RSQLite` package. One step at a time, we can connect like this:

```
library(RSQLite)
driver <- dbDriver("SQLite")
db_file <- system.file("extdata", "crabtag.sqlite", package = "learningr")
conn <- dbConnect(driver, db_file)

query <- "SELECT * FROM Daylog"
head(daylog <- dbGetQuery(conn, query))
##   Tag ID Mission Day      Date Max Temp Min Temp Max Depth Min Depth
## 1 A03401           0 08/08/2008    27.73    25.20      0.06     -0.07
## 2 A03401           1 09/08/2008    25.20    23.86      0.06     -0.07
## 3 A03401           2 10/08/2008    24.02    23.50     -0.07     -0.10
## 4 A03401           3 11/08/2008    26.45    23.28     -0.04     -0.10
## 5 A03401           4 12/08/2008    27.05    23.61     -0.10     -0.26
## 6 A03401           5 13/08/2008    24.62    23.44     -0.04     -0.13
##   Batt Volts
## 1       3.06
## 2       3.09
## 3       3.09
## 4       3.09
## 5       3.09
## 6       3.09

dbDisconnect(conn)

## [1] TRUE

dbUnloadDriver(driver)
```

```
## [1] TRUE
```

Or, more cheekily, we can use the function defined in Chapter 12:

```
head(daylog <- query_crab_tag_db("SELECT * FROM Daylog"))
```

Or we can use the dbReadTable function to simplify things a little further:

```
get_table_from_crab_tag_db <- function(tbl)
{
  driver <- dbDriver("SQLite")
  db_file <- system.file("extdata", "crabtag.sqlite", package = "learningr")
  conn <- dbConnect(driver, db_file)
  on.exit(
    {
      dbDisconnect(conn)
      dbUnloadDriver(driver)
    }
  )
  dbReadTable(conn, tbl)
}
head(daylog <- get_table_from_crab_tag_db("Daylog"))
```

```
##    Tag.ID Mission.Day       Date Max.Temp Min.Temp Max.Depth Min.Depth
## 1 A03401           0 08/08/2008    27.73    25.20      0.06     -0.07
## 2 A03401           1 09/08/2008    25.20    23.86      0.06     -0.07
## 3 A03401           2 10/08/2008    24.02    23.50     -0.07     -0.10
## 4 A03401           3 11/08/2008    26.45    23.28     -0.04     -0.10
## 5 A03401           4 12/08/2008    27.05    23.61     -0.10     -0.26
## 6 A03401           5 13/08/2008    24.62    23.44     -0.04     -0.13
##    Batt.Volts
## 1       3.06
## 2       3.09
## 3       3.09
## 4       3.09
## 5       3.09
## 6       3.09
```

Exercise 13-1

1. To detect question marks, use str_detect:

```
library(stringr)
data(hafu, package = "learningr")
hafu$FathersNationalityIsUncertain <- str_detect(hafu$Father, fixed("?"))
hafu$MothersNationalityIsUncertain <- str_detect(hafu$Mother, fixed("?"))
```

2. To replace those question marks, use str_replace:

```
hafu$Father <- str_replace(hafu$Father, fixed("?"), "")
hafu$Mother <- str_replace(hafu$Mother, fixed("?"), "")
```

Exercise 13-2

Make sure that you have the reshape2 package installed! The trick is to name the measurement variables, "Father" and "Mother":

```
hafu_long <- melt(hafu, measure.vars = c("Father", "Mother"))
```

Exercise 13-3

Using base R, table gives us the counts, and a combination of sort and head allows us to find the most common values. Passing useNA = "always" to table means that NA will always be included in the counts vector:

```
top10 <- function(x)
{
  counts <- table(x, useNA = "always")
  head(sort(counts, decreasing = TRUE), 10)
}
top10(hafu$Mother)

## x
## Japanese       <NA>    English  American    French    German   Russian
##      120         50         29        23        20        12        10
##   Fantasy     Italian  Brazilian
##        8           4          3
```

The plyr package provides an alternative solution that returns the answer as a data frame, which may be more useful for further manipulation of the result:

```
top10_v2 <- function(x)
{
  counts <- count(x)
  head(arrange(counts, desc(freq)), 10)
}
top10_v2(hafu$Mother)

##              x freq
## 1    Japanese  120
## 2        <NA>   50
## 3     English   29
## 4    American   23
## 5      French   20
## 6      German   12
## 7     Russian   10
## 8     Fantasy    8
## 9      Italian    4
## 10  Brazilian    3
```

Exercise 14-1

1. cor calculates correlations, and defaults to Pearson correlations. Experiment with the method argument for other types:

```
with(obama_vs_mccain, cor(Unemployment, Obama))

## [1] 0.2897
```

2. In base/lattice/ggplot2 order, we have:

```
with(obama_vs_mccain, plot(Unemployment, Obama))
xyplot(Obama ~ Unemployment, obama_vs_mccain)
ggplot(obama_vs_mccain, aes(Unemployment, Obama)) +
  geom_point()
```

Exercise 14-2

1. Histograms:

```
plot_numbers <- 1:2
layout(matrix(plot_numbers, ncol = 2, byrow = TRUE))
for(drug_use in c(FALSE, TRUE))
{
  group_data <- subset(alpe_d_huez2, DrugUse == drug_use)
  with(group_data, hist(NumericTime))
}

histogram(~ NumericTime | DrugUse, alpe_d_huez2)

ggplot(alpe_d_huez2, aes(NumericTime)) +
  geom_histogram(binwidth = 2) +
  facet_wrap(~ DrugUse)
```

2. Box plots:

```
boxplot(NumericTime ~ DrugUse, alpe_d_huez2)
bwplot(DrugUse ~ NumericTime, alpe_d_huez2)

ggplot(alpe_d_huez2, aes(DrugUse, NumericTime, group = DrugUse)) +
  geom_boxplot()
```

Exercise 14-3

For simplicity, the answers to this are given using ggplot2. Since this is a data exploration, there is no "right" answer: if the plot shows you something interesting, then it was worth drawing. When you have several "confounding factors" (in this case we have year/ethnicity/gender), there are lots of different possibilities for different plot types. In general, there are two strategies for dealing with many variables: start with an overall summary and add in variables, or start with everything and remove the variables that don't look very interesting. We'll use the second strategy.

The simplest technique for seeing the effects of each variable is to facet by everything:

```
ggplot(gonorrhoea, aes(Age.Group, Rate)) +
  geom_bar(stat = "identity") +
  facet_wrap(~ Year + Ethnicity + Gender)
```

There is a difference between the heights of the bars in each panel, particularly with respect to ethnicity, but with 50 panels it's hard to see what is going on. We can simplify by plotting each year together on the same panel. We use group to state

---

which values belong in the same bar, `fill` to give each bar a different fill color, and `position = "dodge"` to put the bars next to each other rather than stacked one on top of another:

```
ggplot(gonorrhoea, aes(Age.Group, Rate, group = Year, fill = Year)) +
  geom_bar(stat = "identity", position = "dodge") +
  facet_wrap(~ Ethnicity + Gender)
```

The reduced number of panels is better, but I find it hard to get much meaning from all those bars. Since most of the age groups are the same width (five years), we can cheat a little and draw a line plot with age on the x-axis. The plot will be a bit wrong on the righthand side of each panel, because the age groups are wider, but it's close enough to be informative:

```
ggplot(gonorrhoea, aes(Age.Group, Rate, group = Year, color = Year)) +
  geom_line() +
  facet_wrap(~ Ethnicity + Gender)
```

The lines are close together, so there doesn't seem to be much of a time trend at all (though the time period is just five years). Since we have two variables to facet by, we can improve the plot slightly by using `facet_grid` instead of `facet_wrap`:

```
ggplot(gonorrhoea, aes(Age.Group, Rate, group = Year, color = Year)) +
  geom_line() +
  facet_grid(Ethnicity ~ Gender)
```

This clearly shows the effect of ethnicity on gonorrhoea infection rates: the curves are much higher in the "Non-Hispanic Blacks" and "American Indians & Alaskan Natives" groups. Since those groups dominate so much, it's hard to see the effect of gender. By giving each row a different y-scale, it's possible to neutralize the effect of ethnicity and see the difference between males and females more clearly:

```
ggplot(gonorrhoea, aes(Age.Group, Rate, group = Year, color = Year)) +
  geom_line() +
  facet_grid(Ethnicity ~ Gender, scales = "free_y")
```

Here you can see that women have higher infection rates than men (with the same age group and ethnicity) and have a more sustained peak, from 15 to 24 rather than 20 to 24.

Exercise 15-1

1. Both the number of typos in this case, x, and the average number of typos, `lambda`, are 3:

   ```
   dpois(3, 3)
   ```

   ```
   ## [1] 0.224
   ```

2. The quantile, q, is 12 (months), the `size` is 1, since we only need to get pregnant once, and the `probability` each month is 0.25:

   ```
   pnbinom(12, 1, 0.25)
   ```

```
## [1] 0.9762
```

3. Straightforwardly, we just need to set the probability to 0.9:

```
qbirthday(0.9)
```

```
## [1] 41
```

Exercise 15-2

We can either take a subset of the dataset (using the usual indexing method or the subset function), or pass subsetting details to the subset argument of lm:

```
model1 <- lm(
  Rate ~ Year + Age.Group + Ethnicity + Gender,
  gonorrhoea,
  subset = Age.Group %in% c("15 to 19" ,"20 to 24" ,"25 to 29" ,"30 to 34")
)
summary(model1)

##
## Call:
## lm(formula = Rate ~ Year + Age.Group + Ethnicity + Gender, data = gonorrhoea,
##     subset = Age.Group %in% c("15 to 19", "20 to 24", "25 to 29",
##         "30 to 34"))
##
## Residuals:
##    Min     1Q Median     3Q    Max
## -774.0 -127.7  -10.3  106.2  857.7
##
## Coefficients:
##                                   Estimate Std. Error t value Pr(>|t|)
## (Intercept)                        9491.13   25191.00    0.38   0.7068
## Year                                 -4.55      12.54   -0.36   0.7173
## Age.Group20 to 24                   131.12      50.16    2.61   0.0097
## Age.Group25 to 29                  -124.60      50.16   -2.48   0.0138
## Age.Group30 to 34                  -259.83      50.16   -5.18  5.6e-07
## EthnicityAsians & Pacific Islanders -212.76     56.08   -3.79   0.0002
## EthnicityHispanics                 -124.06      56.08   -2.21   0.0281
## EthnicityNon-Hispanic Blacks       1014.35      56.08   18.09  < 2e-16
## EthnicityNon-Hispanic Whites       -174.72      56.08   -3.12   0.0021
## GenderMale                          -83.85      35.47   -2.36   0.0191
##
## (Intercept)
## Year
## Age.Group20 to 24                   **
## Age.Group25 to 29                   *
## Age.Group30 to 34                   ***
## EthnicityAsians & Pacific Islanders ***
## EthnicityHispanics                  *
## EthnicityNon-Hispanic Blacks        ***
## EthnicityNon-Hispanic Whites        **
## GenderMale                          *
## ---
```

```
## Signif. codes:  0 '***' 0.001 '**' 0.01 '*' 0.05 '.' 0.1 ' ' 1
##
## Residual standard error: 251 on 190 degrees of freedom
## Multiple R-squared:  0.798,  Adjusted R-squared:  0.789
## F-statistic: 83.7 on 9 and 190 DF,  p-value: <2e-16
```

Year isn't significant, so we remove it:

```
model2 <- update(model1, ~ . - Year)
summary(model2)

##
## Call:
## lm(formula = Rate ~ Age.Group + Ethnicity + Gender, data = gonorrhoea,
##     subset = Age.Group %in% c("15 to 19", "20 to 24", "25 to 29",
##         "30 to 34"))
##
## Residuals:
##    Min     1Q Median     3Q    Max
## -774.0 -129.3   -6.7  104.3  866.8
##
## Coefficients:
##                                   Estimate Std. Error t value Pr(>|t|)
## (Intercept)                          358.2       53.1    6.75  1.7e-10
## Age.Group20 to 24                    131.1       50.0    2.62  0.00949
## Age.Group25 to 29                   -124.6       50.0   -2.49  0.01363
## Age.Group30 to 34                   -259.8       50.0   -5.19  5.3e-07
## EthnicityAsians & Pacific Islanders -212.8       55.9   -3.80  0.00019
## EthnicityHispanics                  -124.1       55.9   -2.22  0.02777
## EthnicityNon-Hispanic Blacks        1014.3       55.9   18.13  < 2e-16
## EthnicityNon-Hispanic Whites        -174.7       55.9   -3.12  0.00207
## GenderMale                           -83.8       35.4   -2.37  0.01881
##
## (Intercept)                         ***
## Age.Group20 to 24                   **
## Age.Group25 to 29                   *
## Age.Group30 to 34                   ***
## EthnicityAsians & Pacific Islanders ***
## EthnicityHispanics                  *
## EthnicityNon-Hispanic Blacks        ***
## EthnicityNon-Hispanic Whites        **
## GenderMale                          *
## ---
## Signif. codes:  0 '***' 0.001 '**' 0.01 '*' 0.05 '.' 0.1 ' ' 1
##
## Residual standard error: 250 on 191 degrees of freedom
## Multiple R-squared:  0.798,  Adjusted R-squared:  0.79
## F-statistic: 94.5 on 8 and 191 DF,  p-value: <2e-16
```

This time Gender is significant, so we can stop. (Infection rates are similar across genders in children and the elderly, where sex isn't the main method of transmision, but in young adults, women have higher infection rates than men.)

Most of the possible interaction terms aren't very interesting. We get a much better fit by adding an ethnicity-gender interaction, although not all interaction terms are significant (output is verbose, and so not shown). If you're still having fun analyzing, try creating a separate ethnicity indicator for black/non-black and use that as the interaction term with gender:

```
##
## Call:
## lm(formula = Rate ~ Age.Group + Ethnicity + Gender + Ethnicity:Gender,
##     data = gonorrhoea, subset = Age.Group %in% c("15 to 19",
##         "20 to 24", "25 to 29", "30 to 34"))
##
## Residuals:
##    Min    1Q Median    3Q    Max
## -806.0 -119.4   -9.4  105.3  834.8
##
## Coefficients:
##                                                  Estimate Std. Error t value
## (Intercept)                                         405.1       63.9    6.34
## Age.Group20 to 24                                   131.1       50.1    2.62
## Age.Group25 to 29                                  -124.6       50.1   -2.49
## Age.Group30 to 34                                  -259.8       50.1   -5.19
## EthnicityAsians & Pacific Islanders                -296.9       79.2   -3.75
## EthnicityHispanics                                 -197.2       79.2   -2.49
## EthnicityNon-Hispanic Blacks                        999.5       79.2   12.62
## EthnicityNon-Hispanic Whites                       -237.0       79.2   -2.99
## GenderMale                                         -177.6       79.2   -2.24
## EthnicityAsians & Pacific Islanders:GenderMale      168.2      112.0    1.50
## EthnicityHispanics:GenderMale                       146.2      112.0    1.30
## EthnicityNon-Hispanic Blacks:GenderMale              29.7      112.0    0.27
## EthnicityNon-Hispanic Whites:GenderMale             124.6      112.0    1.11
##                                                  Pr(>|t|)
## (Intercept)                                      1.7e-09 ***
## Age.Group20 to 24                                0.00960 **
## Age.Group25 to 29                                0.01377 *
## Age.Group30 to 34                                5.6e-07 ***
## EthnicityAsians & Pacific Islanders              0.00024 ***
## EthnicityHispanics                               0.01370 *
## EthnicityNon-Hispanic Blacks                     < 2e-16 ***
## EthnicityNon-Hispanic Whites                     0.00315 **
## GenderMale                                       0.02615 *
## EthnicityAsians & Pacific Islanders:GenderMale   0.13487
## EthnicityHispanics:GenderMale                    0.19361
## EthnicityNon-Hispanic Blacks:GenderMale          0.79092
## EthnicityNon-Hispanic Whites:GenderMale          0.26754
## ---
## Signif. codes:  0 '***' 0.001 '**' 0.01 '*' 0.05 '.' 0.1 ' ' 1
##
## Residual standard error: 251 on 187 degrees of freedom
## Multiple R-squared:  0.802,  Adjusted R-squared:  0.789
## F-statistic: 63.2 on 12 and 187 DF,  p-value: <2e-16
```

Exercise 15-3

First, the setup. Get the package installed in the usual way:

```
install.packages("betareg")
```

We need to rescale the response variable to be between 0 and 1 (rather than 0 and 100), because that is the range of the beta distribution:

```
ovm <- within(obama_vs_mccain, Obama <- Obama / 100)
```

Now we're ready to run the model. It's just the same as running a linear regression, but we call betareg instead of lm. I'm arbitrarily looking at Black as the ethnicity group and Protestant as the religion:

```
library(betareg)
beta_model1 <- betareg(
  Obama ~ Turnout + Population + Unemployment + Urbanization + Black + Protestant,
  ovm,
  subset = State != "District of Columbia"
)
summary(beta_model1)

##
## Call:
## betareg(formula = Obama ~ Turnout + Population + Unemployment +
##     Urbanization + Black + Protestant, data = ovm, subset = State !=
##     "District of Columbia")
##
## Standardized weighted residuals 2:
##    Min    1Q Median    3Q    Max
## -2.834 -0.457 -0.062  0.771  2.044
##
## Coefficients (mean model with logit link):
##               Estimate Std. Error z value Pr(>|z|)
## (Intercept)  -5.52e-01   4.99e-01   -1.11   0.2689
## Turnout       1.69e-02   5.91e-03    2.86   0.0042 **
## Population    1.66e-09   5.34e-09    0.31   0.7566
## Unemployment  5.74e-02   2.31e-02    2.48   0.0132 *
## Urbanization -1.82e-05   1.54e-04   -0.12   0.9059
## Black         8.18e-03   4.27e-03    1.91   0.0558 .
## Protestant   -1.78e-02   3.17e-03   -5.62  1.9e-08 ***
##
## Phi coefficients (precision model with identity link):
##       Estimate Std. Error z value Pr(>|z|)
## (phi)    109.5       23.2    4.71  2.5e-06 ***
## ---
## Signif. codes:  0 '***' 0.001 '**' 0.01 '*' 0.05 '.' 0.1 ' ' 1
##
## Type of estimator: ML (maximum likelihood)
## Log-likelihood: 72.1 on 8 Df
## Pseudo R-squared: 0.723
## Number of iterations: 36 (BFGS) + 3 (Fisher scoring)
```

Urbanization is nonsignificant, so we remove it, using the same technique as for lm:

```
beta_model2 <- update(beta_model1, ~ . - Urbanization)
summary(beta_model2)

##
## Call:
## betareg(formula = Obama ~ Turnout + Population + Unemployment +
##     Black + Protestant, data = ovm, subset = State != "District of Columbia")
##
## Standardized weighted residuals 2:
##    Min    1Q Median    3Q    Max
## -2.831 -0.457 -0.053  0.774  2.007
##
## Coefficients (mean model with logit link):
##              Estimate Std. Error z value Pr(>|z|)
## (Intercept) -5.69e-01   4.77e-01   -1.19   0.2327
## Turnout      1.70e-02   5.86e-03    2.90   0.0037 **
## Population   1.73e-09   5.31e-09    0.32   0.7452
## Unemployment 5.70e-02   2.29e-02    2.48   0.0130 *
## Black        7.93e-03   3.73e-03    2.13   0.0334 *
## Protestant  -1.76e-02   2.48e-03   -7.09  1.3e-12 ***
##
## Phi coefficients (precision model with identity link):
##       Estimate Std. Error z value Pr(>|z|)
## (phi)    109.4       23.2    4.71  2.5e-06 ***
## ---
## Signif. codes:  0 '***' 0.001 '**' 0.01 '*' 0.05 '.' 0.1 ' ' 1
##
## Type of estimator: ML (maximum likelihood)
## Log-likelihood: 72.1 on 7 Df
## Pseudo R-squared: 0.723
## Number of iterations: 31 (BFGS) + 3 (Fisher scoring)
```

Similarly, Population has to go:

```
beta_model3 <- update(beta_model2, ~ . - Population)
summary(beta_model3)

##
## Call:
## betareg(formula = Obama ~ Turnout + Unemployment + Black + Protestant,
##     data = ovm, subset = State != "District of Columbia")
##
## Standardized weighted residuals 2:
##    Min    1Q Median    3Q    Max
## -2.828 -0.458  0.043  0.742  1.935
##
## Coefficients (mean model with logit link):
##              Estimate Std. Error z value Pr(>|z|)
## (Intercept) -0.55577    0.47567   -1.17   0.2427
## Turnout      0.01686    0.00585    2.88   0.0040 **
## Unemployment 0.05964    0.02145    2.78   0.0054 **
## Black        0.00820    0.00364    2.26   0.0241 *
```

```
## Protestant    -0.01779    0.00240   -7.42  1.2e-13 ***
##
## Phi coefficients (precision model with identity link):
##        Estimate Std. Error z value Pr(>|z|)
## (phi)    109.2        23.2    4.71 2.5e-06 ***
## ---
## Signif. codes:  0 '***' 0.001 '**' 0.01 '*' 0.05 '.' 0.1 ' ' 1
##
## Type of estimator: ML (maximum likelihood)
## Log-likelihood: 72.1 on 6 Df
## Pseudo R-squared: 0.723
## Number of iterations: 27 (BFGS) + 2 (Fisher scoring)
```

Plotting can be done with exactly the same code as with `lm`:

```
plot_numbers <- 1:6
layout(matrix(plot_numbers, ncol = 2, byrow = TRUE))
plot(beta_model3, plot_numbers)
```

Exercise 16-1

There is no prescription for the exact rules that must be followed to check and handle user input. In general, if input is in the wrong form, it is best to try to convert it to the right one, warning the user in the process. Thus, for nonnumeric inputs, we can try to convert them to be numeric with a warning.

For nonpositive values, we can try substituting an in-range value or pretend the value was missing, but this would make the answer wrong. In this case, it is best to throw an error:

```
harmonic_mean <- function(x, na.rm = FALSE)
{
  if(!is.numeric(x))
  {
    warning("Coercing 'x' to be numeric.")
    x <- as.numeric(x)
  }
  if(any(!is.na(x) & x <= 0))
  {
    stop("'x' contains non-positive values")
  }
  1 / mean(1 / x, na.rm = na.rm)
}
```

Exercise 16-2

Here's the RUnit version. To test missing values, we need to consider the cases with `na.rm` being TRUE and `na.rm` being FALSE. To test that a warning has been thrown, we use the trick of changing the warn option to 2, meaning that warnings are treated as errors. For testing nonpositive numbers, we use the boundary case of zero:

```
test.harmonic_mean.1_2_4.returns_12_over_7 <- function()
{
  expected <- 12 / 7
```

```
    actual <- harmonic_mean(c(1, 2, 4))
    checkEqualsNumeric(expected, actual)
}
test.harmonic_mean.no_inputs.throws_error <- function()
{
    checkException(harmonic_mean())
}
test.harmonic_mean.some_missing.returns_na <- function()
{
    expected <- NA_real_
    actual <- harmonic_mean(c(1, 2, 4, NA))
    checkEqualsNumeric(expected, actual)
}
test.harmonic_mean.some_missing_with_nas_removed.returns_12_over_7 <- function()
{
    expected <- 12 / 7
    actual <- harmonic_mean(c(1, 2, 4, NA), na.rm = TRUE)
    checkEqualsNumeric(expected, actual)
}
test.harmonic_mean.non_numeric_input.throws_warning <- function()
{
    old_ops <- options(warn = 2)
    on.exit(options(old_ops))
    checkException(harmonic_mean("1"))
}
test.harmonic_mean.zero_inputs.throws_error <- function()
{
    checkException(harmonic_mean(0))
}
```

The testthat translation is straightforward. Warnings are easier to test for, using the expect_warning function:

```
expect_equal(12 /7, harmonic_mean(c(1, 2, 4)))
expect_error(harmonic_mean())
expect_equal(NA_real_, harmonic_mean(c(1, 2, 4, NA)))
expect_equal(12 /7, harmonic_mean(c(1, 2, 4, NA), na.rm = TRUE))
expect_warning(harmonic_mean("1"))
expect_error(harmonic_mean(0))
```

Exercise 16-3

Here's the updated harmonic_mean function:

```
harmonic_mean <- function(x, na.rm = FALSE)
{
    if(!is.numeric(x))
    {
        warning("Coercing 'x' to be numeric.")
        x <- as.numeric(x)
    }
    if(any(!is.na(x) & x <= 0))
    {
        stop("'x' contains non-positive values")
```

```
    }
    result <- 1 / mean(1 / x, na.rm = na.rm)
    class(result) <- "harmonic"
    result
  }
```

To make an S3 method, the name must be of the form function.class; in this case, print.harmonic. The contents can be generated with other print functions, but here we use the lower-level cat function:

```
print.harmonic <- function(x, ...)
{
  cat("The harmonic mean is", x, "\n", ...)
}
```

Exercise 17-1

Creating the function and the data frame is straightforward enough:

```
sum_of_squares <- function(n)
{
  n * (n + 1) * (2 * n + 1) / 6
}
x <- 1:10
squares_data <- data.frame(x = x, y = sum_of_squares(x))
```

In your call to package.skeleton, you need to think about which directory you want to create the package in:

```
package.skeleton("squares", c("sum_of_squares", "squares_data"))
```

Exercise 17-2

The function documentation goes in *R/sum_of_squares.R*, or similar:

```
#' Sum of Squares
#'
#' Calculates the sum of squares of the first \code{n} natural numbers.
#'
#' @param n A positive integer.
#' @return An integer equal to the sum of the squares of the first \code{n}
#' natural numbers.
#' @author Richie Cotton.
#' @seealso \code{\link[base]{sum}}
#' @examples
#' sum_of_squares(1:20)
#' @keywords misc
#' @export
```

The package documentation goes in *R/squares-package.R*:

```
#' squares: Sums of squares.
#'
#' A test package that contains data on the sum of squares of natural
#' numbers, and a function to calculate more values.
#'
```

```
#' @author Richie Cotton
#' @docType package
#' @name squares
#' @aliases squares squares-package
#' @keywords package
NULL
```

The data documentation goes in the same file, or in *R/squares-data.R*:

```
#' Sum of squares dataset
#'
#' The sum of squares of natural numbers.
#' \itemize{
#'   \item{x}{Natural numbers.}
#'   \item{y}{The sum of squares from 1 to \code{x}.}
#' }
#'
#' @docType data
#' @keywords datasets
#' @name squares_data
#' @usage data(squares_data)
#' @format A data frame with 10 rows and 2 variables.
NULL
```

Exercise 17-3

The code is easy; the hard part is fixing any problems:

```
check("squares")
build("squares")
```

Bibliography

1. Jason R. Briggs. *Python for Kids: A Playful Introduction to Programming*. 2012. William Pollock. ISBN-13 978-1-59327-407-8.

2. Garrett Grolemund. *Data Analysis with R*. 2013. O'Reilly. ISBN-13 978-1-4493-5901-0.

3. Andrie de Vries and Joris Meys. *R For Dummies*. 2012. John Wiley & Sons. ISBN-13 978-1-1199-6284-7.

4. Michael Fitzgerald. *Introducing Regular Expressions*. 2012. O'Reilly. ISBN-13 978-1-4493-9268-0.

5. Paul Murrell. *R Graphics*, Second Edition. 2011. Chapman and Hall/CRC. ISBN-13 978-1-4398-3176-2.

6. Hadley Wickham. *ggplot2: Elegant Graphics for Data Analysis*. 2010. Springer. ISBN-13 978-0-3879-8140-6.

7. Deepayan Sarkar. *Lattice: Multivariate Data Visualization with R*. 2008. Springer. ISBN-13 978-0-3877-5968-5.

8. Edward R. Tufte. *Envisioning Information*. 1990. Graphics Press USA. ISBN-13 978-0-9613-9211-6.

9. Michael J. Crawley. *The R Book*. 2013. John Wiley & Sons. ISBN-13 978-0-4709-7392-9.

10. Andy Field, Jeremy Miles, and Zoe Field. *Discovering Statistics Using R*. 2012. SAGE Publications. ISBN-13 978-1-4462-0046-9.

11. Max Kuhn. *Applied Predictive Modeling*. 2013. Springer. ISBN-13 978-1-4614-6848-6.

12. John Fox and Sanford Weisberg. *An R Companion to Applied Regression*. 2011. SAGE Publications. ISBN-13 978-1-4129-7514-8.

13. José Pinheiro and Douglas Bates. *Mixed-Effects Models in S and S-PLUS*. 2009. Springer. ISBN-13 978-1-4419-0317-4.

14. Graham Williams. *Data Mining with Rattle and R: The Art of Excavating Data for Knowledge Discovery*. 2011. Springer. ISBN-13 978-1-4419-9889-7.

15. Thomas Lumley. "Standard nonstandard evaluation rules". 2003. *http://developer.r-project.org/nonstandard-eval.pdf*.

16. Dirk Eddelbuettel. *Seamless R and C++ Integration with Rcpp*. 2013. Springer. ISBN-13 978-1-4614-6867-7.

17. Yihui Xie. *Dynamic Documents with R and knitr*. 2013. Chapman and Hall/CRC. ISBN-13 978-1-4822-0353-0.

18. Michael Lawrence and John Verzani. *Programming Graphical User Interfaces in R*. 2012. Chapman and Hall/CRC. ISBN-13 978-1-4398-5682-6.

19. Q. Ethan McCallum and Stephen Weston. *Parallel R*. 2012. O'Reilly. ISBN-13 978-1-4493-0992-3.

Index

Symbols

! operator, 21
# symbol, 8
$ operator, 62
% operator, 96
%% operator, 15
%*% operator, 53
%/% operator, 15
%in% operator, 82
%o% operator, 53
& operator, 21
+ operator, 13, 300
: operator, 13, 13, 39
<- operator, 18, 306
<<- operator, 18, 306
== operator, 15
? operator, 8
?? operator, 8, 10
[[]] indexing, 62
[] indexing, 43, 61
^ operator, 53
| operator, 21

A

adding dates, 160
adding vectors, 13
aggregate function (stats), 137
alarm function (utils), 98

all.equal function (base), 16
anova function (stats), 273
apply function (base), 192
*apply function family, 138
apropos function (utils), 9
Architect, 6 (see Eclipse)
args function (base), 84
arguments, 8
    formal, 83
    pass into functions, 70
    using functions as, 87
arithmetic
    date/time, 160
    lists, 60
    on arrays, 52–54
arrange function (plyr), 201
array function (base), 46
arrays, 46–54
    arithmetic on, 52–54
    columns of, 48–51
    creating, 46
    dimensions of, 48–51
    indexing, 51
    looping over, 132–135
    rows of, 48–51
    two-dimensional, 29
as* functions (base, methods), 32
assertive package, 31
assign function (base), 18, 80, 300

We'd like to hear your suggestions for improving our indexes. Send email to index@oreilly.com.

ggplot2 graphics package
    bar charts in, 258–260
    box plots in, 252
    histograms in, 247–248
    line plots in, 233–237
    scatterplots in, 224–230
GitHub, 148
glm, 280
global assignment, 18, 306
global environment, 79, 81
global error option, 293
global variables, 89
gnumeric package, 181
Gnumeric spreadsheets, 181
Google Code, 148
graphical representations, 207–261
    bar charts, 253–260
    histograms, 238–248
    line plots, 230–237
    scatterplots, 212–230
    summary statistics, 207–211
grep function (base), 192
grepl function (base), 192
GROUP BY operations (SQL), 138
gWidgets framework, 331

H

HDF5 files, 181
head function (utils), 34
    internal nodes and, 176
help function (utils), 9
help search operator, 10
help, sources for, 8–11
help.search function (utils), 9
hist function (graphics), 238
histograms, 238–248
    in base graphics, 238–244
    in ggplot2 graphics, 247–248
    in lattice graphics, 244–247
HTML files, 175
htmlParse function (XML), 185

I

IDEs, 5–7
    Architect, 6
    Eclipse, 6
    Emacs, 5
    ESS, 5

Live-R, 7
    popular choices, 5
    Revolution-R, 7
    RStudio, 6
    XEmacs, 6
if, 300
if/else statements, 112
    vectorized, 114
ifelse function (base), 113, 114
Ihaka, Ross, 3
importing
    XML files, 175
    YAML files, 178
%in% operator, 82
INDEX files, 314
indexing, 43
    arrays, 51
    data frames, 74–75
    elements (vectors), 43–45
    lists, 61–64
infinite values, 19
initialize method, 306
inst directory, 314
install.packages function (utils), 148
install.Rtools function (installr), 313
installed.packages function (utils), 146
installr package, 11, 313
install_github function (devtools), 150
integer class, 26
Integrated Development Environment (see IDEs)
integration testing, 294
interpreter, 299–302
intervals, 163
inverse CDF, 266
is.* function family (base, methods), 30

J

JDBC API, 185
JGR, 7
JSON files, 176–179

K

knitr package, 315

L

labelling elements, 42

About the Author

Richard Cotton is a data scientist with a background in chemical health and safety, and has worked extensively on tools to give nontechnical users access to statistical models. He is the author of the R packages `assertive` (for checking the state of your variables) and `sig` (to make sure your functions have a sensible API). He runs The Damned Liars statistics consultancy.

Colophon

The animals on the cover of *Learning R* are roe deer (*Capreolus capreolus*), a species of deer found throughout much of Europe, Scandinavia, and the Mediterranean region. *Roe* is derived from the Old English word for "spotted," though other translations show that it might be an ancient word for "red."

The roe deer is rather small (averaging 3–4 feet long and around 2 feet tall at the shoulders), with long graceful legs. Male roe have short antlers with just a few branches, but are otherwise nearly the same size as female roe. These deer also have very short tails with a white rump patch, which they flash when alarmed by something. In the summer, their coats are red, but fade to gray or pale brown in the winter. Their fawns are born in summer after a 10-month gestation period, and have white spots for their first six weeks of life.

Woodland is the preferred habitat of the roe deer, though they also graze in grasslands and sparse forests. Occasionally they can be found near farms, but tend to avoid fields where livestock have been kept, perhaps because the grass is trampled and less clean. They eat grass, shoots, leaves, and berries, and are most active at twilight.

In the Austrian books upon which the classic Disney film *Bambi* was based, Bambi was a roe deer—but Disney changed the character to a white-tailed deer, which was more familiar to North American audiences.

The cover image is from Cassell's *Natural History*. The cover font is Adobe ITC Garamond. The text font is Adobe Minion Pro; the heading font is Adobe Myriad Condensed; and the code font is Dalton Maag's Ubuntu Mono.

O'REILLY®

There's much more where this came from.

Experience books, videos, live online training courses, and more from O'Reilly and our 200+ partners—all in one place.

Learn more at oreilly.com/online-learning

CPSIA information can be obtained
at www.ICGtesting.com
Printed in the USA
BVHW052017100520
579414BV00011B/355